Between the Moon and Earth

A Scientific Exploration of Earth-Based Heavens and Hells

JAY ALFRED

Order this book online at www.trafford.com
or email orders@trafford.com

Most Trafford titles are also available at major online book retailers.

Print information available on the last page.

ISBN: 978-1-4120-9505-1 (sc)
ISBN: 978-1-6987-0330-5 (e)

Cover Photograph: Separate images of the Earth and Moon, which were taken in 1992 by the Galileo spacecraft, were combined to generate this view. Courtesy of NASA/JPL-Caltech

Trafford rev. 12/02/2022

www.trafford.com
North America & international
toll-free: 844-688-6899 (USA & Canada)
fax: 812 355 4082

Contents

Other Books By Jay Alfred

Our Invisible Bodies
Brains and Realities

Introduction

Since the dawn of history, various cultures have imagined heaven and the abode of gods to be literally located in some part of the sky; and hell, to be in the interior of the Earth. With recent developments in science, this may not be as far-fetched as once thought. From a modern scientific perspective, these were the earliest concepts of parallel realties and the idea of a multiverse. The idea of a multiverse is a serious area of research in cutting-edge physics today. Some, like Bernard Carr, professor of mathematics and astronomy at Queen Mary University of London, believe (like the author) that scientific theories should be expanded and extended to include credible metaphysical observations. The model presented here, which gives a modern understanding of heavens and hells, is supported by scientific evidence and correlates to science-based metaphysical observations. It uses, as its basis, the physics of dark (i.e., invisible) matter, as well as plasma physics and photonics.

About 85 per cent of the matter in our universe is composed of dark matter – a mysterious form of invisible matter that scientists are still trying to specifically identify. However, there is an enormous amount of evidence, from various sources, that dark matter exists. World-renowned theoretical physicist and cosmologist, Stephen Hawking, summarizes:

> Our galaxy and other galaxies must contain a large amount of "dark matter" that we cannot see directly, but which we know must be there because of the influence of its gravitational attraction on the orbits of stars in the galaxies.
>
> Stephen Hawking, Theoretical Physicist

It is interesting that more than 99 per cent of our ordinary matter universe is in the state of plasma, i.e., non-atomic matter. The author believes that dark matter is composed of two components. The first makes up the bulk of dark matter and is essentially an invisible pervasive field of light of slow-moving and neutral particles, where dark photonics (or the study of dark photons or light particles) is important. The second smaller component is found only in the vicinity of ordinary matter, and is in the state of a dark plasma, i.e., a plasma of charged dark matter particles (the details are discussed in chapter 7,) where plasma physics is important.

After the death of the ordinary matter body, the author believes that most human beings will find themselves in a higher energy sphere, within what scientists would now call a "dark sector," which interpenetrates and shares the same spacetime landscape and gravitational field as the visible, ordinary matter Earth. They will move about using high-energy, light, and resilient, dark plasma bodies. Depending on the properties of their bodies they will gravitate or levitate into particular ecological niches within the shells in the relevant sphere — higher energy shells being coincident with Earth's atmosphere and beyond, and lower energy shells being coincident with the surface of the Earth and below. This book discusses the locations of these niches, or colloquially called "heavens" and "hells," relative to the ordinary matter Earth.

Focus of the Book

While the nature of dark plasma bodies has been discussed in detail in the author's book "Our Invisible Bodies," this book expands on the environments in which these bodies inhabit, including the nature of the Earth-based heavens and hells, that was first introduced in the previous book. The focus in this book will therefore be on the nature and structure of these environments within the Earth-Moon system, and their possible locations, using a scientific framework.

The Journey

We will begin with a review of heavens and hells in religion and history, and then do a quick review of Earth's structure, before introducing the reader to the nature of dark matter, plasma, and photonics. This will give you the background knowledge to follow the thought process in building a model and a mental map to locate the traditional heavens and hells within the Earth-Moon system. (There will be no math involved.) We will then explore some interactions between our world and these parallel realities. The book will end by looking at how the dark infrastructure of the Earth-Moon system is connected to the rest of the Solar System, the universe, and the local multiverse. A glossary at the end of the book includes a list of the scientific and metaphysical terms used and their meanings.

Jay Alfred
May 2006, November 2022

Every soul is ordained to wander
between incarnations in the region
between the Moon and Earth.

Plutarch,
Ancient Greek Philosopher

CHAPTER 1

❋

What's the Weather in Heaven Today?

Earth-Based Heavens in Religion and History

Charles Leadbeater, the twentieth-century experimental metaphysicist, says that the average person, passing into heaven-life, tends to float at a considerable distance above the surface of the Earth. Hiralal Kaji, a medium, transmits that the planes that most people go to, after the death of their ordinary matter bodies, are in the atmosphere.

> While climbing, we leave the world behind and move in an upward direction. The region which the soul has to traverse is our atmosphere.
>
> Hiralal Kaji, channeling a spirit

This is also stated or implied in the world's religious scriptures:

> From where the sun arises to where it sets, there all the gods are suspended... as god he dwells in the atmosphere.
>
> Katha Upanishad, Hindu Scriptures

> For whom in the sky is comparable to Yahweh?
>
> Psalms, Jewish Old Testament

> As they were watching, he [i.e., Jesus] was lifted up, and a cloud took him out of their sight.
>
> The Acts of the Apostles,
> Christian New Testament

Edward Wright, associate professor of "Hebrew Bible and Early Judaism" at the University of Arizona, notes that ancient Egyptians can be credited with having one of the earliest concepts of human beings having an afterlife in the heavenly realms. They believed that heaven was a physical place far above the Earth. Departed souls are believed to undergo a literal journey to reach heaven, along the way being blocked by hazards and entities who try to deny the soul from reaching heaven. The peoples of Mesopotamia and elsewhere, he says, also imagined that there was a realm "up there" populated by divine beings.

The writings attributed to Socrates, and the "Myth of Er," written in the fourth century BC, by the philosopher, Plato, was probably one of the most influential pieces of literature that established the notions of an immortal soul, heavens, and hells – which influenced early Christianity greatly. It is interesting to note that, in Greek, the same word is used for both "heaven" and "sky" i.e., "ouranos." The same is true in Hebrew and Aramaic, one word for both sky and heaven. The two were not clearly distinguished by the ancients. Many centuries later, the popular medieval view of heaven was that it existed as a physical place above the clouds and that God, and the angels, were physically above, watching over man.

Different Levels of Heaven

During biblical times, the Jews did not have the scientific knowledge that we now take for granted. Instead, they tried to conceptualize the world in terms of their intuitive knowledge, and usually described it visually. So, when they conceived of the universe, they constructed a multi-layered world, sort of like a large onion composed of various layers with the physical world in which human beings lived at the center. These layers were called "firmament" or "shamayim" (heavens or sky) in the Old Testament or "heavens" in the New Testament era.

A common feature about heavens in many cultures is that there were many "levels." (Saint) Paul writes in the Christian New Testament: "I know a man in Christ who, fourteen years ago, whether in the body or out of the body I do not know...was caught up to the third heaven." A "third heaven" raises the idea that there are heavens at various levels. There was a widespread belief in the region around the Mediterranean and the Middle East, before the fall of the Western Roman Empire in AD 476, that there were different levels of heavens. The Quran and Hadeeth of the Muslims contained seven levels of heavens. In other texts, sometimes, even ten levels were identified. Although the number of levels or layers was different in the various models, it is interesting that practically all the major religions in the world speak of different levels of heavens, above the surface of the Earth.

In the Buddhist cosmography, six levels of gods live on or above the central mountain, Mount Meru. "Bodhisattvas" or pre-Buddhas live in "Tusita Heaven" before they descend to the human realm. Buddhists believe that the historical Buddha (i.e., Siddhartha Gautama) dwelt there, before he descended to be born in India. They believe that Maitreya, "the future Buddha," is now living there; and is awaiting the time when he will descend and enter the human realm. Christians believe that Jesus descended from heaven and after his resurrection ascended back to it.

The Divine Comedy, a poem which described an imaginary journey through hell, purgatory, and paradise, was written by Dante Alighieri in the 1300s. In Canto 1 of the poem, after an initial ascension, Dante is guided through nine concentric spheres (or levels) of heavens. This model was heavily influenced by Aristotelian and Ptolemaic cosmography. They are, however, not astronomical but metaphysical heavens.

"Astronomical" vs. Metaphysical Heavens

The term "heavens" was also used in early astronomy. These are not "metaphysical heavens" (as discussed above) – they housed observable astronomical bodies. The most influential treatise in history, relating to these types of "astronomical heavens," was Aristotle's Earth-centered cosmological work "On the Heavens," written in the fourth century BC. Aristotle proposed that the heavens were literally composed of fifty-five concentric, crystalline spheres to which observable celestial objects were attached, and which rotated at different velocities, with the Earth at the center. Ptolemy, too, in the second century AD, had devised an elaborate model of the heavens - composed of large and small circles, with the Earth in the center. This was later superseded by the Sun-centered model of Copernicus in the sixteenth century.

Conclusion

What all these beliefs and theories have in common is that there were many different levels of heavens juxtaposed against the atmosphere and beyond, of the visible Earth. Historically, the visible Earth was conflated with the invisible counterpart spheres that metaphysics talked about. Theories of concentric spherical rings or shells of heavens, which included celestial

CHAPTER 2

*

Where the Hell am I?

Earth-Based Hells In Religion and History

Just like culture and technology, concepts of the afterlife, including hell, evolved, and transformed through the centuries. Even, within a specific culture, these ideas can be quite different at different points of time – whether ancient, medieval, or modern. Although many hells involved torture and suffering, some just required annihilation of the evil (without any suffering,) while others were just places where the dead, both good and evil, would go. In most concepts of hells, the torture and suffering is temporary and is meant to be rehabilitative and based on restorative justice. It is only eternal in the mainstream interpretation of the later Christian and Islamic concepts of hell. In certain religions, like Buddhism and Hinduism, everyone will be ultimately saved from suffering after long eons of rebirths, culminating in enlightenment. However, in mainstream Christianity only a small minority will. The majority will be tortured for eternity. Hence, there are many different concepts of the nature of hell. One thing they all had in common, though, was that hell was thought to be located below the surface of the Earth.

Where is Hell?

The ancient Mesopotamians believed hell lay underground, only a short distance from the Earth's surface. The underworld also figures prominently in Egyptian mythology. The pharaohs were reputed to be in contact with the gods under Earth's surface and visited them regularly, using a system of secret tunnels in the pyramids. Ancient Egyptians, however, did not have any hell where souls suffered or were tortured. The evil souls were simply annihilated, although the good souls enjoyed an eternal heaven. The Greeks believed that the dead (both good and evil) descended to Hades, an enormous cavern deep below the ocean, utterly devoid of sunlight. There was a terrifying area, however, within Hades, where evil souls faced endless torture and punishment, which was called Tartarus.

Buddhist cosmography identifies "four planes of deprivation," which belong to the animal, ghost, and demi-gods realms, as well as one to "hell." (Within this realm there are many different levels and types of hells.) In Buddhism one is not thrown into hell by anyone or punished by some deity. Rather, the hells are the fruition of one's misdeeds and the externalization of one's own character. The hells of Vedic cosmology would also be situated here. In China, the dead (both good and evil) were believed to descend to a murky subterranean land, known as Huang-quan. However, this was not eternal, due the operation of karma. Ancient Taoism, however, had no concept of hell, as morality was seen to be an evolving and arbitrary human manufactured construct. Later, it was influenced by Buddhism. The Catholic version of hell as a place was confirmed at Fatima in 1917, during the church-approved apparition of "Our Lady of Fatima" (believed to be Mary, the mother of the historical Jesus) to three young shepherd children. Lucia Santos, the eldest of three children, reported a vision of hell in 1941, as follows, "Our Lady showed us a great sea of fire which seemed to be *under the Earth*" (emphasis added.)

Evolution of the Christian Hell

Bart Ehrman is a leading authority on the New Testament and the history of early Christianity and a Distinguished Professor of Religious Studies at the University of North Carolina at Chapel Hill. He notes that the early Jewish did not actually have any developed concept of an afterlife. The soul is just like the breath, which ceases on death. The ancient Hebrews believed that some undefined shadowy aspect of the dead descended deep below ground, to a netherworld known as "She'ol." However, this was not hell as we know it, as both good and evil people went there. There were also no elaborate schemes of torture or suffering.

Subsequently, about two centuries before the time of Jesus, to provide some sense of justice with regards to the fate of good and evil people, the Jews introduced the concept of a final judgement day, when the (ordinary matter) bodies of everyone will be resurrected. On this day, the good souls will be rewarded with the joys of heaven and the evil would simply be annihilated. Annihilation was considered an "eternal" punishment as there would be no reversal (i.e., they would not be resurrected at a later date.) This was the belief of the Jews during the time of Jesus, and of early Christians. It was strongly influenced by the Egyptian belief in the resurrection of the physical body. This was why in Egypt mummification and other preservative methods were required of the dead ordinary matter body. It was believed that the person's "spiritual double" (i.e., the physical-etheric body, which will be discussed in later chapters) would rejoin and transform the physical body at some future date and live again. This is almost identical to the later Christian belief.

Later on, in the first century, when more non-Jews, or Hellenistic Gentiles, became Christians, new ideas were infused into Christianity. Greek and Hellenistic ideas were especially influential in the "civilized world" at this time. The newly converted Gentiles then grafted Christianity with the idea of an immortal soul that leaves the body on death and goes to heaven or hell. These ideas came mainly from Plato's "Myth of Er," which was found at the end of his work called "The Republic," around the fourth century BC. This evolved into the early Christian and medieval conception of hell as being associated with fire: "everlasting fire," "unquenchable fire," and "flame." It is a place of torment with "fire and brimstone" and a "lake of fire and brimstone" where the wicked are "tormented day and night." Brimstone refers to sulfur, which is often found in volcanic areas which leads deep into the Earth. The descriptions fit well with the conditions in the hot interior of the Earth.

Different Levels of Hell

While Siddhartha Gautama (the historical and most well-known "Buddha" or enlightened one) rejected the divine authority of the Vedas and many of the key assumptions of Brahmanism, he affirmed, on the basis of his own direct knowledge, the existence of many of the elements of the Vedic world view, including the existence of numerous hells, heavens, and various supernatural beings like gods, demons, and ghosts. He not only claimed to have seen these realms and beings for himself with his "divine sight," but also how sentient beings cycle through these many diverse forms of existence. Buddhist and Vedic cosmography shows a variety of hells at distinct levels. There are eight hot hells and eight cold hells, within the hell realm, which lie thousands of kilometers beneath the surface of the "southern continent." The hells are stacked one on top of the other, with the worst hell at the bottom.

In Dante's time, in the middle ages in Europe, it was believed that hell existed underneath Jerusalem. His cosmography and infernal geography clearly show various levels of hell. In his famous poem the "Divine Comedy," Dante divided hell into nine concentric shells of increasing nastiness, with the worst hell right at the bottom. Each concentric shell represented further and further evil, culminating in a frozen lake in the center of the Earth.

Towards a Scientific Understanding of Heavens and Hells

It doesn't seem likely, based on current theories of Earth's geological composition, that an ordinary matter hell is below the crust of the Earth. Geologists and seismologists have constructed quite a detailed picture of the Earth's interior by probing its core with seismic shockwaves from earthquakes and nuclear tests. The geological facts rule out any notions of undiscovered places in Earth's interior where conscious physical beings (composed of ordinary matter) could inhabit.

Many of the descriptions of heavens and hells in the historical literature make clear references to locations in the ordinary matter Earth. However, they do not make a distinction between the ordinary matter Earth and parallel realities. Hence, they sometimes combined landscapes, beings, and objects in the ordinary matter universe with parallel universes. This betrays the notion that both physical and "super-physical" spheres interpenetrated each other. In certain experiences, they were actually superimposed on each other. This will not be surprising if we had multiple bodies, with different cognitive-sensory systems generating overlapping streams of consciousness.

In order to begin a deeper analysis, it may be useful to make a *quick* tour of the Earth, from its core to the outer reaches of its atmosphere and beyond, in the next chapter. This will provide a clear understanding of the locations at which heavens and hells were situated in the ordinary matter Earth, based on the various historical and religious accounts. It will also help us gain some understanding about where approximately we should posit the various heaven and hells, after obtaining a modern scientific perspective on the possible nature of these heavens and hells, using dark matter and plasma physics (there will be no math involved.)

CHAPTER 3

❋

Quick Tour of the Earth

The ordinary matter Earth, as we know it, is a three dimensional near-sphere. It has a radius of approximately 6,500 km or 4,040 mi. Science has identified concentric layers or shells in the ordinary matter Earth, which will be discussed below, starting from the core in the interior and then moving outwards to the atmosphere and beyond.

From the Core to the Crust

Most of the information that scientists have of the *lower* shells in the ordinary matter Earth, below the surface, comes from the way seismic shockwaves are distorted and reflected through the planet during earthquakes, the activity of the geomagnetic field, and the composition of meteorites from space. From this evidence, it has been established that the ordinary matter Earth (below the surface) is composed of three principal layers: the crust, the mantle, and the core. If you imagine a near-spherical egg, the yolk would be the core, the white albumen would be the mantle above it; and the shell would be the crust covering all of it. Although the core and mantle are about equal in thickness, the core actually forms only 15 percent of the Earth's volume, while the mantle occupies 84 percent. The crust makes up the remaining 1 percent.

The Inner Core

At the center of the ordinary matter Earth is a solid metal core of iron and nickel, a sphere with a radius of some 1,280 km (795 mi.) The boundary of this sphere is 5,100 km (3,170 mi) from the surface. At this depth, the pressure is tremendous and is estimated to be over 14.2 million times the atmospheric pressure. The inner core is separated from the rest of the planet by the molten outer core (discussed below.) Recent estimates by the University of Cambridge say that it rotates in the same direction as the Earth but slower than the entire Earth. The temperature of the inner core is hotter than the Sun's surface. Though this temperature is sufficiently hot to melt iron, the inner core remains solid because of the immense pressure exerted by the matter above it. The solid inner core grows as material from the liquid outer core solidifies on its surface. This continually releases heat to the outer core.

The Outer Core

The outer core is 2,210 km (1,373 mi) thick and is a river of hot liquid metals, namely iron and nickel, circulating around the inner core. The heat released during the solidification of the inner core drives the convection motion of the fluid in the outer core. The motion brings with it electrically conductive material which generates Earth's magnetic field, which shields us from harmful solar radiation.

The Mantle

The Earth's entire mantle, a region of hot rock that, over vast, geological timescales, churns like a fluid, is some 2,855 km (1,775 mi) thick. The fluid is highly viscous — more viscous than the molten outer core. It's made of solid rock but behaves like a hyper-viscous liquid over long time-frames. In a sense, it is really between solid (like the inner core) and liquid (like the outer core.)

The Crust

Although it feels solid and hard beneath our feet, the outer surface of the Earth is a thin crust of fragile rock, fractured like the cracked shell of an egg and "floating" on the mantle. It is composed of granite and basalt rock up to 35 km (22 mi) thick — although it varies from place to place. The pieces of the broken shell are Earth's tectonic plates which float across a layer of soft rock, their motions being driven by forces generated deep within the Earth. At their boundaries, the plates spread apart, converge, and slide past one another. After the crust, on the surface of the Earth, we come to the atmosphere.

The Atmosphere and Beyond

The Earth is surrounded by a blanket of air, which we call the atmosphere. The atmosphere has no abrupt cut-off point. It slowly becomes thinner and fades away into the vacuum of space. For practical purposes, however, the

"Karman line," which is 100 km (62 mi) from the Earth's surface, is often used as the boundary between Earth's atmosphere and space. More than 99 per cent of the mass of atmosphere is below this line.

Lower Atmosphere - *The Troposphere*

The troposphere is the lowest layer of the atmosphere starting at the surface going up to between 7 km (4 mi) at the poles and 17 km (10.5 mi) at the equator with some variation due to weather factors. This part of the atmosphere is the densest. As you climb higher in this layer, the temperature drops from about 17 to -52 degrees Celsius. The air pressure at the top of the troposphere is only 10 per cent of that at sea level. The troposphere holds 99 per cent of the water vapor in the atmosphere. Almost all weather is in this region. Scientists call this layer the lower atmosphere.

Middle Atmosphere - *The Stratosphere and Mesosphere*

The stratosphere starts just above the troposphere and extends to 50 km (31 mi) high. Compared to the troposphere, this part of the atmosphere is dry and less dense. The temperature in this region increases gradually to -3 degrees Celsius, due to the absorption of ultraviolet radiation from the Sun. The ozone layer, which absorbs and scatters the ultraviolet radiation, is in this layer. 99 per cent of the air is found in the troposphere and stratosphere. The next layer, the mesosphere, starts just above the stratosphere and extends to between 80 km (50 mi) and 85 km (53 mi) high. In this region, the temperature decreases to as low as -93 degrees Celsius. Scientists call these layers the middle atmosphere.

Upper Atmosphere - *The Thermosphere and Ionosphere*

The thermosphere starts above the mesosphere and just below the "Karman line" (the "official" boundary of the atmosphere) but extends beyond it to 600 km (373 mi) high. The temperatures go up as you increase in altitude due to the Sun's energy, even as high as 1,727 degrees Celsius. Chemical reactions occur much faster here than on the surface of the Earth. The ionosphere (which overlaps with the thermosphere) is where many atoms are ionised (i.e., they are pushed into a plasma state through collisions with the most energetic sub-atomic particles from the Sun.) Although radio waves pass readily through walls, they are reflected back by this zone due to the electromagnetic properties of plasma, making long-distance radio communications possible. This is also where auroras take place. The ionosphere is the closest of several electrified sheaths that surround and protect the Earth. Scientists call these layers the upper atmosphere.

The Exosphere

The exosphere starts at the top of the thermosphere and continues for almost 190,000 km (118,000 mi) until it merges with interplanetary gases and the vacuum of space. In this region, hydrogen and helium are the main components and are only present at extremely low densities.

Concentric Shells

Although scientists have suffixed the various layers as "spheres" this is, strictly speaking, not very accurate. For example, the "troposphere" is not a sphere in the sense that a basketball would be, as it does not extend to the center of the Earth, In fact, it surrounds the Earth, and it sits on top of the Earth's crust. It is in fact a layer, a *ring* (from a two-dimensional perspective,) or a *shell* (from a three-dimensional perspective) around the Earth. From the above descriptions of the various layers, it is easy to see that Earth's three-dimensional sphere is organized into *concentric shells* around the inner core. If we cut a cross-section, passing through the center of the Earth, the resulting section would look like rings in a bull's eye pattern – with the densest layer in the center and the most tenuous in the outer reaches.

Earth's Magnetic Field

Interpenetrating Earth's atmosphere and extending far into space is the geomagnetic field, which is tilted by about eleven degrees to Earth's spinning axis. As discussed above, it is generated by the activity in the outer core. It rotates with the outer core and its invisible grid lines can vary in intensity from place to place, based on variations in gravitational force, the presence or absence of large mineral deposits like quartz, and the presence of underground streams or large aquifers, all of which can alter the magnetic field on the Earth's surface. Animals, including birds and turtles, can detect the Earth's magnetic field, and use it to navigate during migrations. Cows and wild deer align their bodies in the north-south direction of the field while relaxing, but not when they are under high voltage power lines. This suggest that Earth's magnetism is responsible

Earth's Magnetosphere

The magnetic field creates and controls a region called the magnetosphere that surrounds Earth. With the blast of the solar wind, travelling at supersonic speeds, the magnetosphere is flattened on the day side (i.e., the side facing the Sun) and a tail forms at the night side (the side facing away from the Sun,) giving it a tear-drop shape. The magnetosphere shelters the surface of the planet, like a cloak, from high energy particles in the solar wind which is harmful to life-forms. At the lower limit, the magnetosphere ends at the ionosphere in the upper atmosphere.

Earth's Plasmasphere

Within the magnetosphere is the plasmasphere, which extends out t0 1 to 1.5 Earth diameters, and during quiet conditions up to 3 Earth diameters. Due to space weather activities and gravitational interactions, the plasmasphere never has exact boundaries. It expands and contracts according to these conditions. Since it is within the magnetosphere, it is a region composed of *magnetized* plasma. The plasmasphere is filled with cold plasma from the Earth's ionosphere, hot plasma from the Sun's outer atmosphere, and even hotter plasma accelerated to great speeds which "rains" down on the upper atmosphere causing the auroras in the northern and southern hemispheres. Just like the magnetosphere, the shape of the plasmasphere is a "tear-drop" shape, with the side facing the Sun flattened by the solar wind, and a long-tailed pointed end facing away from the Sun. It is enclosed in a "plasma sheath," whose size and shape is determined by the voltage difference between the planet and the nearby solar plasma.

The Moon's 3d Ordinary Matter Sphere

The Moon has a radius of 1,750 km (1,088 mi.) This means it is a little more than one quarter the radius of the Earth and only about half the radius of Earth's core. Its volume is only about 2 per cent of the Earth's. The Moon has no atmosphere or magnetic field. Scientists say that it has a crust which is about 70 km (43 mi) thick, a mantle about 1,350 km (840 mi) thick, and a core that is about 330 km (205 mi) in radius. A limited amount of seismic data suggests that the outer core may be molten. Although there is a small amount of geological activity on the Moon, it is largely "geologically dead." Most lunar seismic activity appears to be triggered by tidal forces induced in the Moon by Earth's gravitational forces. These forces also cause the Moon to become slightly distorted in shape. Hence, it is not a perfect sphere.

Conclusion

Having gained an understanding of the ordinary matter Earth's structure, we will now review the latest scientific perspectives on dark matter and plasma physics. Putting it altogether, we will then realize and perhaps be surprised to learn that many of the historical accounts of heavens and hells may, after all, have some factual basis, although conflated historically with the visible ordinary matter Earth.

CHAPTER 4

❇

Can Someone Turn On the Lights?

> It's pretty amazing that after all this time; astronomers cannot say what this dark matter is made of. It's one of the greatest mysteries in the history of science... Imagine living in a house and having no clue as to what it is made of.
>
> Tom Siegfried, Editor-in-Chief, Science News

Missing Mass

Strange things are happening. Stars at the edge of galaxies are moving at much higher velocities than allowed by Newton's law of universal gravitation. In our Solar System, the movement of planets occurs in close conformity with Newton's laws, with planets further out from the Sun moving slower. When this behavior is extrapolated to distant spiral galaxies, it is natural to assume that stars further away from the galactic centers would also move more slowly.

It was therefore a surprise when measurements by the astronomer Jan Oort in the late 1920s showed that orbital velocities of stars in the Milky Way do not decrease with increasing distance from our galactic center. In 1933, Fritz Zwicky noted the same anomaly in galaxies orbiting around the center of mass of galactic clusters and suggested that it was due to unidentified invisible "dark matter" which provided the missing mass to account for the additional gravity.

This critical insight by Zwicky was almost forgotten until, in the 1970s, Vera Rubin and Kent Ford, of the Carnegie Institution of Washington, were confused by their observations of the Andromeda Galaxy. The stars at the edges were moving just as fast as those near the center. Subsequently, the same observation was made for over sixty other spiral galaxies! This shows that gravity from visible galaxies extended much further out than what was suggested by the edges of the visible galaxies. But what was generating this gravity?

Dark Galaxies and Stars

A deep optical image, from the telescope at Kitt Peak, showed the cluster of galaxies, Abell 2218, along with many faint blue galaxies in the distant background which had been distorted by gravity into arcs in the images constructed. According to astronomers, these types of distortions were the result of "gravitational lensing," usually caused by a high density of matter near the center of the cluster. Gravitational lensing is when gravity bends or distorts light in a similar way that a lens does. In this case, while gravitational lensing was evident, the matter which caused it could not be found by astronomers! They were forced to conclude that invisible galaxies were causing the lensing phenomenon. A team of European astronomers noticed similar distortions in light from distant bright galaxies they were imaging. But there was no visible object that could account for the distortions. This is not the only case. A group of isolated galaxies — UGC 10214, has a conspicuous bridge of material extending into space towards apparently nothing!

In 2000 Robert Minchin and his team at Cardiff University in Wales noticed two apparently isolated hydrogen clouds in a radio telescope survey of the Virgo Cluster of galaxies. Follow-up observations with visible-light telescopes showed that one of these clouds was associated with a faintly glowing galaxy. However, the second cloud had no partner galaxy. According to Minchin, its motion suggested that it's a small part of a massive object weighing as much as a galaxy of one hundred billion suns, and yet this object was invisible. Minchin and his team found the first member of a population of galaxies that theorists have proposed but observers had never seen – a dark galaxy (i.e., an invisible galaxy with hardly any visible ordinary matter.) Numerous dark galaxies have subsequently been found, with some scientists concluding that there could be more dark galaxies than ordinary matter galaxies.

There were also stars that appeared to be rotating around invisible dark companions. Physicist David Peat explains that while the dark matter star remains invisible, its attraction will affect the orbit of the ordinarily visible star. The two will form a binary system in which a star in our universe is rotating around an invisible companion.

What's Lurking in the Dark?

All this evidence made many scientists sit up and conclude, much to their disbelief, that 99 per cent of the mass (matter and energy) in the universe was dark or invisible (including baryonic i.e., ordinary dark matter, such as black holes, brown dwarfs, and invisible hot gas)! Bruce Margon, chairperson of the astronomy department at the University of Washington, told the New York Times, "It's a fairly embarrassing situation to admit that we can't find [more than] 90 percent of the universe!"! Equally ironic is that although this fact was known to scientists for some decades, 99 per cent of human beings on this planet currently are probably unaware of this historical monumental finding and have not reflected on the implications!

This invisible matter was dubbed "dark matter" by scientists, not because it is dark in color (or because there was something sinister or evil about it,) but because it does not radiate any light within the electromagnetic spectrum measurable by our (ordinary matter) scientific instruments. It could just as well be called "invisible matter." Astronomers are still not sure what exactly this dark matter is. They have labored on several theories but none of this invisible matter predicted by these theories have actually been observed or measured directly by current scientific instruments. Dark matter remains invisible over the entire electromagnetic spectrum, known currently to science, from the longest wave-length radio waves to the shortest wave-length gamma rays. Though astronomers have opened one spectral window after another, with new satellite observatories sent above the absorbing effects of our atmosphere, most of the dark matter consists of matter that cannot be directly detected.

Cosmic Web of Dark Matter

In 2007, an international team of astronomers using NASA's Hubble Space Telescope created the first three-dimensional map of the large-scale distribution of dark matter in the universe. What they found was what looked like a network or web of dark matter filaments, collapsing under the intensive pull of gravity, and growing clumpier and more compact over time. This is generally referred to as the "cosmic dark matter web." Obviously, we cannot see this dark matter web directly, but how do we know it exists?

A team led by astrophysicist Norbert Werner of the Netherlands Institute for Space Research (NISR) in Utrecht identified a huge filament of hot gas (composed of ordinary matter particles) stretching between two clusters of galaxies, Abell 222 and Abell 223, located about 50 million light-years apart, and 2.3 billion light-years away. The filament is thought to be one thread in a vast web of filaments. Computer simulations have been telling us for several years that most of the "missing" gas in the universe should be in hot filaments," said Smita Mathur, an Ohio State University researcher. "Most of those filaments are too faint to see, but it looks like we are finally finding their shadows." Scientists have now confirmed that these filaments of hot gas exist. This web of huge hot filaments of ordinary matter particles, spanning long distances and connecting stars, galaxies, and clusters of galaxies, betrays the presence of an underlying cosmic dark matter web that is providing the gravitational scaffolding for these filaments.

Theoretical Predictions

In 1932 Albert Einstein and William de Sitter published a joint paper in which they proposed the Einstein-de Sitter model of the universe. They predicted in this paper that there might be large amounts of matter which does not emit light and has not been detected. Today, powerful scientific theories imply that the greater part of this dark matter is composed of "exotic" particles not yet seen in the laboratory. For e.g., supersymmetry and superstring theories predict whole sets of new particles that are not in the standard model of particle physics (this is discussed in more detail in the next chapter.) Astronomers have also recently calculated that the mass so far observed in the visible universe fell far short of theoretical predictions of the established "inflationary theory" of standard cosmology. The results from various sources show that the observed density of matter and energy in the universe was simply too low – if only ordinary matter was accounted for. There is therefore a widespread consensus now in the scientific community that massive amounts of dark matter pervade the universe. Scientists today estimate that exotic dark matter forms almost 85 per cent of all matter in the universe. In the Milky Way, it is even higher, at 95 per cent.

Dark matter within the Milky Way

For decades astronomers have wondered about the origin of certain fast-moving clouds of atomic hydrogen in the vicinity of the Milky Way. Some clouds appeared to be plunging into the galaxy at high speeds and were not rotating with the galaxy, while others seemed to be moving away from it. Leo Blitz of the University of California and David Spergel of Princeton University say that these high velocity clouds might harbor dark matter — a hypothesis which would account for the continued stability of the clouds and their unexplained large internal velocities.

D Lin, a University of California astronomer, has shown that the Large Magellanic Cloud that orbits around our galaxy is being torn apart by the powerful gravitational pull of a dense cloud of dark matter surrounding the Milky Way. This dismemberment of the Large Magellanic Cloud cannot be explained by the gravitational forces exerted by the visible stars in our galaxy. New observations in 2021 show that the cloud has created a wake, like a ship sailing through calm waters, as it travels through the Milky Way's halo. Lin calculates that our galaxy's halo of dark matter is equivalent to 600 to 800 billion solar masses, compared to the only 100 billion solar masses of visible matter. Overall, as evidenced by fast-moving stars at the edge, the galaxy sits inside a huge dark matter halo, with a shape approximating an ovoid.

Dark Matter within the Sun

According to researchers from the University of Oxford (as reported in the "New Scientist" journal,) the Sun is harboring a vast reservoir of dark matter. Astrophysicists Ilidio Lopes and Joe Silk reasoned that passing dark matter particles would be captured by the gravity of heavy bodies like the Sun. Not surprisingly, the source of high energy super-physical particles, as identified by metaphysicists, such as qi by the Chinese, and prana by the Indians, is also said to be the Sun.

Metaphysical Evidence for Dark Matter Particles from the Sun

Charles Leadbeater, a leading metaphysicist of the twentieth century, and Barbara Brennan, a research scientist and astrophysicist who worked for NASA and now a subtle energy healer, say that there are high-energy charged globules or bundles of vibrating energy that are abundant and energetic on sunny days. On cloudy days, though, they move sluggishly and diminish in numbers. Brennan has also observed them being absorbed by plant life. These energetic globules are considered one type of "qi" by the Chinese or "prana" by the Indians — both examples of what metaphysicists call "subtle matter and energy."

If more of these globules are received on Earth during sunny days than on cloudy days does this suggest that they emanate from our Sun? Axions, one type of dark matter particle, is thought by scientific researchers to emanate from the Sun. Since axions can convert to ordinary photons in a magnetic field and back again (including possibly Earth's magnetic field under certain conditions,) it would be possible for them to be blocked by clouds after converting to ordinary photons. There could be a variety of dark matter particles from the Sun, with different properties.

H P Blavatsky, a leading metaphysicist, said in the nineteenth century, "the Sun is the storehouse of vital force, and that from the Sun issue those life-*currents* which thrill through space, as through the organisms of every living thing on Earth — the real Sun being hidden behind the visible Sun and generating the vital *fluid* which circulates throughout our (Solar) System in a ten year cycle." It is significant that she mentions the real Sun being hidden behind the visible Sun. It is an obvious reference to a Sun which is ordinarily invisible to most of us and our scientific instruments — a dark matter Sun!

In addition to heat and light, the Sun constantly emits a low density plasma of charged electrons and protons called the "solar wind," which blasts out from the Sun in all directions at exceedingly high speeds to fill the entire Solar System and beyond. The solar wind is in fact a plasma of charged particles. Plasma can act as a *fluid*. Furthermore, moving charged particles are *currents*. (When Blavatsky wrote this, the word "plasma" had yet to be invented in 1929, and the notion of a solar wind was not yet established. British astrophysicist Arthur Eddington suggested the existence of the solar wind only in 1910.) The composition of this wind has been largely analyzed by science to consist of only ordinary matter in the state of a plasma. However, since dark matter particles like axions and dark photons can oscillate back and forth into ordinary photons, they may be carried by the solar wind (as ordinary photons which may convert back to dark matter particles.)

Furthermore, if there is a dark matter reservoir in the Sun (what Blavatsky characterised as "a storehouse of vital force,) self-interacting charged dark matter particles (discussed in the next chapter) could also be blown out as a dark plasma wind or be accelerated as a weak dark current flowing along the dark magnetic field lines of Earth's dark matter counterparts (discussed in chapter 13,) originating from the dark matter Sun, i.e., using the same transport mechanism that creates auroras. These are the currents that "thrill through space," according to Blavatsky, who has identified them as the lowest energy dark matter particles, known as the "physical-etheric" particles in the metaphysical literature. The currents are known in science as Birkeland currents and are discussed in chapter 7.

It states in the metaphysical literature (including Taoist and Hindu literature) that qi or prana emanates from the Sun. The Sun and other stars have significant and pervasive magnetic fields and plasma winds and currents that blow or send out both ordinary and dark matter particles from their hot interiors. Standing in front of the Sun would be like standing in front of an electric fan blowing out qi or prana. Qigong practitioners therefore face the Sun and other stars during their practice. In Chinese culture (under the study of "feng-shui,") it is "auspicious" for the door of a Chinese home to face the east (in the direction of the sunrise.) Qi and prana are described as "subtle" energy (or radiant matter) by metaphysicists because it interacts weakly with ordinary matter.

Dark Matter within the Earth-Moon System

Dark Matter Clouds and Rain

Jürg Diemand, a physicist at the University of California in Santa Cruz, US, and colleagues say that new computations suggest that small clouds of dark matter, which could be detected by future space missions, pass through

Earth on a regular basis. He says that perhaps a million billion of them drift around the large dark matter halo that is thought to enclose our galaxy. These clouds float through Earth every ten thousand years in an encounter lasting for about fifty years. However, they do not affect the ordinary matter Earth to any appreciable effect. Their relatively low densities mean they could only nudge our planet out of its normal orbit by less than a millionth of a meter per second.

Astrophysicist Heidi Newberg at Rensselaer Polytechnic Institute and her colleagues suggest that dark matter may be raining down on Earth from the dwarf galaxy "Sagittarius." The dwarf galaxy is being torn apart and consumed by the much larger gravitational pull of our galaxy, the Milky Way. It's entrails of stars and dust which forms a tail, like a long piece of ribbon, is entangled around and within our galaxy. This tail extends from Sagittarius' center and then arcs across and below the plane of the Milky Way. The leading part of the tail extends northward above our galaxy where it then turns and appears to be showering shredded galaxy debris down directly on our Solar System.

As the Milky Way consumes Sagittarius, it not only rips the stars from the smaller galaxy, but also tears away some of the dark matter particles from that galaxy. We may be able to directly observe that in the form of a dark matter highway streaming in one direction through the Earth," says Newberg, who has recently identified stars near the Sun that could be part of this tail. Professor Rubia of the Italian National agency for new technologies, energy, and environment (ENEA,) speaking at the 2004 Institute of Physics Nuclear Physics Conference in Edinburgh, UK, says that a stream of dark matter particles might constantly be flowing through the Earth, and these could be measurable with underground detectors.

Dark Seasons on Earth

The Earth is constantly being bombarded by dark matter particles as it, and the Solar System, sweep around the dark matter halo of the Milky Way, well beyond supersonic speeds. The Sun is moving around in the galaxy, and the Earth is moving around the Sun. Consequently, all of us, and any detector we built in the laboratory, is moving through a cloud of the lightest dark matter particles. As Earth orbits around the center of the galaxy, the planet flies through this cloud of dark matter. As that happens, millions of weakly interacting dark matter particles would be raining down on Earth. According to research scientists, there would be an annual modulation or seasonal variation in the amount of dark matter particles raining down on Earth because of the motion of the Earth, relative to the Solar System.

In December, the Earth is moving against the direction of the motion of the Solar System, which is moving around the galactic center. However, in June, it moves in the same direction as the Solar System. This rain of dark matter particles would therefore be experienced seasonally, just like the monsoon rains. In the same way that a cyclist gets wetter in the rain, when riding against the wind's direction, than when riding with the wind, any dark matter detector would record more dark matter particles in and around June than around December. However, the tilt of the Earth's axis could influence the peak season for the "dark matter rain" relative to each country.

It is curious that the Chinese celebrate an ancient festival called the "Hungry Ghosts" month around August each year. They believe that souls are freed in this month to roam the Earth. This belief probably arose from observations that there were more manifestations or sightings of ghosts around this month. If ghosts are composed of dark matter (as suggested in this book and analyzed in more detail in chapter 20) and there is a higher density of dark matter particles during this month, the number of sightings would in fact be higher. The timing of the ancient Chinese festival could therefore have a deeper scientific significance.

Conclusion

The evidence, as a whole, appears to support the existence of invisible dark matter in the universe, including our Solar System, and on Earth. The majority of physicists are now inclined to take the view that there are vast amounts of matter that we cannot see or directly measure (with any scientific instrument that we currently have.) Nevertheless, they knew it was there because of the indirect effects on what they could see — unusually fast moving stars at the edge of spiral galaxies, distortions in the light from distant galaxies indicating invisible mass, and ordinary matter being drawn towards apparently nothing. Other evidence include data from the CMB (Cosmic Microwave Background.) This huge amount of observational evidence, from diverse sources, is supported by rigorous mathematical theories that suggest that there is a huge population of new particles yet to be found. But what exactly are these particles that make up dark matter?

CHAPTER 5

What Exactly is Dark Matter?

Dark Matter Particles

Since exotic dark matter makes up about 85 per cent of all the matter in the universe, it would be simplistic to assume that it would be composed of only one type of particle. There is probably a great diversity of particles included in dark matter, including exotic particles which escape the imagination of both physicists and metaphysicists at present. There are well thought-out theories in physics which shed some light on these particles.

These include dark photons, which mediate a dark electromagnetic force. Axions are light, slow-moving, force-mediating particles, which are also good dark matter candidates. They may mediate a conjectured attractive fifth force between ordinary and dark matter. If so, they could be considered a "cross-interacting" particle i.e., a particle that interacts with both ordinary and dark matter, like gravitons. Both dark photons and axions are particles that mediate forces. They can convert to ordinary photons and back again. According to the latest theories, dark matter particles could also include self-interacting dark protons and electrons. Candidates for dark matter particles include large sets of supersymmetric and mirror particles. Metaphysicists have traditionally called these "superphysical" particles.

Shortcomings of the Standard Model of Particle Physics

Physicists have been using the very successful "standard model of particle physics" for several decades to classify the plethora of elementary particles discovered in the twentieth century. Using this model, physicists were able to speculate on the existence of new particles and later discover them in particle accelerators. This was done by observing the "missing spaces" in the model, just as the "periodic table" was used in an earlier century, in a similar way, to identify and discover new elements. One glaring omission of the standard model is that it does not include the graviton. Since gravitational fields are pervasive through-out the universe, and the graviton is the particle mediating the gravitational force, this is a significant shortcoming of the model.

The limitations of the model forced physicists to think about more comprehensive models, such as those developed by "superstring theories." At low energies, superstring theories were able to derive all the particles under the standard model simply by observing mathematical symmetries. They were also able to derive the graviton. These theories, therefore, seem to provide a more comprehensive and elegant model using supersymmetry. Supersymmetry predicts a whole new set of particles. In supersymmetry theory, standard *matter* particles in the standard model were reflected (as in a mirror) as a new set of "shadow" *force* particles. Conversely, standard *force* particles were reflected as a new set of "shadow" *matter* particles. Current scientific instruments in particle accelerators and various dark matter particle detectors, however, have not yet found any of these new sets of shadow particles at low-energies — although the presence of large amounts of dark matter suggests their presence. Why have they not been observed in the laboratory or any particle accelerators?

Perhaps these invisible particles were highly energetic or massive. It will then take much more energy to create such particles in our current particle accelerators, and they would also decay much faster into other particles in our low-energy ordinary matter universe. However, even with the higher energies that our current accelerators are capable of, we have not seen any of these particles. The author proposes that it could be that, apart from a matter-force symmetry, and a higher energy level, these particles were also mirroring other physical attributes (such as charge and parity.)

Mirror particles are another set of particles that has been identified as possible dark matter candidates. In this case, we use a mirror to reflect the left-handed particles in our ordinary matter universe into a set of right-handed particles in a mirror universe. This is called a "parity-reversal." When mirror particles meet ordinary particles, they interact only weakly. Similarly, we can note that anti-matter particles are conceptually mirror reflected particles of matter with the charge reversed. However, they have not been found in the required abundance at cosmic scales. Could anti-matter, mirror, and supersymmetric particles, all be reflected using a single mirror? In this case, both anti-matter and supersymmetric particles would escape detection if they were also parity-reversed. The dark mirror universe (identified as the physical-etheric universe by metaphysicists) seems to suggest so.

This would provide a solution to the problem of missing anti-matter in the universe at cosmic scales. As anti-matter is also reversed in terms of parity and matter-force symmetry, it will only interact weakly with ordinary matter. (It could also be reversed in spacetime; this will be discussed in more detail in chapter 11.) We will therefore not expect any annihilations of dark matter particles as its anti-particles would be sequestered in a mirror universe. It may be because of this more complicated scenario, where there are multiple reversals of a range of physical attributes using a single mirror, which has probably not been considered in dark matter experimental designs, which is making the detection of dark matter particles, so far, unsuccessful.

"Super-Physical" Particles in Metaphysical Literature

Early in the twentieth century many references to invisible particles, beyond ordinary matter particles, have been made in the writings of Leadbeater and Besant, and in the compilations of Arthur Powell. They would not know about supersymmetry or dark matter and have generally referred to these particles as "super-physical" particles, which means invisible particles, usually beyond the energy levels of visible ordinary matter particles. For a modern interpretation and the usage in this book, we could say super-physical or "super particles" are exotic particles that are beyond the standard model of particle physics. The particles in the current standard model will be called "standard particles."

Theories of invisible particles and interpenetrating universes may be new to mainstream physics but there are detailed references to such particles and universes in metaphysical, including Vedic, Yogic and Buddhist literature. Several millenniums ago, in Hindu metaphysics, we find references to not only "anu" (the standard particles) but also "param-anu" (which means, "beyond the anu," i.e., super particles.)

I K Taimni, a metaphysicist and a professor of chemistry at the Allahabad University in India, discussed at length about the existence of different types of particles in parallel universes. Theoretically "an infinite number of systems with vastly varying sizes of particles and wavelengths can be accommodated in the cosmos," according to him.

> The differences between the sizes of particles and wavelengths can be enormous and so one system can remain within another system without any common ground between the two, and therefore, without the inhabitants of one system knowing of the existence of the other inhabitants.
>
> I K Taimni, 1974, Metaphysicist
> and Professor of Chemistry

He highlighted in the 1970s that there are exotic particles belonging to other universes, which would be ordinarily invisible to us — exactly what scientists are saying now. He explained that the systems or universes will not interfere or come in contact with each other on the "material plane" (i.e., our ordinary matter universe) and their inhabitants. These inhabitants will not be able to recognise the phenomena or inhabitants of other universes or dark sectors as their cognitive and sensory apparatus have evolved in their universes for particular ranges of vibrations (i.e., frequencies) and/or interactions.

Experimental metaphysicist, Leadbeater, discusses a number of times, in various books and papers, about the existence of super-physical and higher-energy particles and interpenetrating universes. These particles, which were counterparts or partners to ordinary matter particles (similar to how it is being conceived in modern supersymmetry theory today,) were given names such as "physical-etheric," "astral," or "mental" particles. There were other names given to even higher energy particles.

In 1913, Leadbeater explained that if the whole of the physical (ordinary matter) brain is spread out so as to be one particle thick, and then the corresponding astral and mental matter, we would have three layers of differing particle densities, all corresponding one to the other, but not joined in any way except here and there with "wires of communication." The observation that ordinary matter particles are linked or correlated to higher energy super-physical particles indicates that there must be some affinity between the particles, Hence, there could be some, as yet unidentified, fundamental force, that weakly binds these particles.

References to the physical-etheric, astral, and other higher energy counterpart particles are clear references to ordinarily invisible super particles. It is an accepted fact in the metaphysical literature that "astral" particles are particles with much higher energies than standard particles. The literature also indicates that there was awareness of the interactions and relationships between standard and super particles. Unlike modern scientists, though, these metaphysicists were able to see the objects composed of these particles, through training, when observing from linked higher-energy dark plasma bodies, but they did not have a rigorous mathematical theory to support it. These bodies are discussed in detail in the author's book "Our Invisible Bodies."

Leadbeater and Besant explained in their book "Occult Chemistry," published in 1919, that a physical atom cannot be directly split into astral atoms. If the force within the "ultimate physical atom" (or "anu") is pushed over the threshold of the astral plane (in other words, if the energy levels are increased beyond the physical plane) "the atom disappears" they say. This force (or more accurately the energy) working on a higher plane then expresses itself through astral atoms. It is significant that they reiterate that the higher energetic atoms "may vanish from the plane" — making it part of invisible (dark) matter. Similar descriptions are given for interactions relating to mirror dark matter particles today. Furthermore, we know that in today's modern particle accelerators new particles can be created at higher energies. The standard particle was referred to as the "physical atom" and the super-particle the "astral atom." The fact that super particles (just like dark matter particles) are invisible from our everyday frame of reference was reiterated many times by various metaphysicists.

Leadbeater says that when a man picks up (let us say) a piece of stone, he sees only the physical particles of that stone, but that in no way affects the "undoubted fact" that that stone at the same time is made up of particles of matter of the physical-etheric, astral, and other higher planes. Once again, we have clear references to super particles, which are often cited in metaphysical literature. According to superstring theory, the super particle is a shadow partner or counterpart of the standard particle. Here, a metaphysicist says the same thing in the nineteenth century, even before atoms were known to exist, and before the theoretical discoveries of mainstream science in the twentieth century:

> The [ordinarily invisible] etheric double is the exact duplicate of the visible body — its shadow, as it were, particle for particle.
>
> Annie Besant, 1896, Metaphysicist

Self-interacting Dark Matter (SIDM)

The media, when reporting about dark matter, often say that dark matter particles are not self-interacting. These views, however, are now slowly being fine-tuned as numerous scientific papers, that the public are generally unaware of, are proposing the idea of a new species of self-interacting dark matter particles in response to puzzling astrophysical observations within galaxies, in the vicinity of ordinary matter. The metaphysical evidence also suggests that subtle dark matter particles do self-interact and seem to project a weak electromagnetic force (see the author's book "Our Invisible Bodies" for references and discussion.) We can call this different species of dark matter particles, which are usually found in the vicinity of ordinary matter, self-interacting dark matter (or SIDM) particles.

Since these dark matter particles seem to shadow ordinary matter and share similar locations, it is proposed that there are equal amounts of ordinary, and dark, self-interacting matter. Since ordinary matter makes up 15 per cent of all matter in the universe, we estimate that self-interacting dark matter also makes up 15 per cent. The rest, making up 70 per cent of the matter in the universe, will consist of dark matter particles that have properties that are already widely accepted by mainstream science, i.e., they are widely dispersed, slow-moving, non-self-interacting, neutral, non-atomic matter.

Conclusion

There is a large amount of anecdotal evidence in the metaphysical literature that there are multiple levels of dark matter particles. Based on this literature, we can conclude that the lowest energy "physical-etheric" particles make up the first level of dark matter particles. There are other particles, already identified by experimental metaphysicists. These particles, at increasing energy levels, are generally called the astral, mental, and spiritual particles. Despite these odd non-scientific names, they fulfill the properties of self-interacting dark matter particles.

Considering that exotic dark matter makes up a whopping 85 per cent of all matter in the universe, with only 15 per cent consisting of standard particles, it is not difficult to imagine more than one level of dark matter particles. On the contrary, it is difficult to imagine numerous standard particles, making up only a small fraction of the matter in the universe, and the bulk being made up of only one type of dark matter particle. Furthermore, supersymmetry theory and mirror particles theory predict large ensembles of new particles, comparable in number to the particles in the standard model.

Where do these super particles live? Shadow and mirror dark matter particles would live in shadow and mirror universes. The next chapter throws light on these universes.

CHAPTER 6

❋

Shadow and Mirror Universes

Shadow Universes in Science

E8xE8 Superstring Theory

According to the E8xE8 superstring theory, a shadow universe interpenetrates our universe. Astrophysicist John Gribbin explains that there are two "E8" components in the theory. However, only one of the E8 components in the theory is required to describe everything in our ordinary matter universe. This means the other E8 component leaves a complete duplicate set of particles and forces. The symmetry between the two E8 components broke at the birth of the universe, when gravity split apart from the other forces of nature in the first second of the big bang. The result was the development of two duplicate universes, each mirroring and interpenetrating the other, but interacting only through gravity. Each universe would have its own complete set of fundamental forces and elementary particles.

The shadow matter from this duplicate universe would neither be seen nor felt. However, the shadow matter particles could be self-interacting, i.e., they could interact with each other. This could give rise to shadow planets.

> There would be shadow photons, shadow atoms, perhaps coexisting in the same spacetime that we inhabit, but forever invisible. A shadow planet could pass right through the Earth and never affect us, except through its gravitational pull.
>
> John Gribbin, Physicist

Physicist Richard Morris elaborates that there could be even shadow matter life-forms.

> If someone tried to grasp a chunk of shadow matter, her hands would pass right through it. It has been said that one would walk through a shadow matter mountain or stand at the bottom of a shadow matter ocean and never know it. Shadow matter particles could interact with one another according to physical laws similar to those of our world. It is possible that there could be shadow matter planets, and perhaps even shadow matter organisms.
>
> Richard Morris, Physicist

Physicist David Peat says that both E8 universes coexist within a single spacetime that is curved by their combined mass and energy. Since the forces (except for the gravitational force) are confined within each group, the elementary particles in one group are invisible when viewed from the other group. There would be a different type of light in this shadow universe which would not directly interact with the ordinary matter particles in our universe, so we will not be able to see it.

> It is indeed theoretically possible for a shadow universe to exist in parallel to our own. While we would feel its gravitational effects, this shadow universe would be otherwise invisible. Photons [light] from the shadow group would have no interaction with the matter in our universe.
>
> David Peat, Physicist

Astrophysicist John Gribbin notes that one reason that the idea (of a shadow universe) has been taken seriously is that there is astronomical and cosmological evidence that a lot of the universe exists in the form of dark matter, detectable gravitationally but not seen. Physicist David Peat agrees. Current theories of the large scale structure of the universe, based on astrophysical measurements, predict that the mass of the universe is much more than what is observed. The idea of a shadow universe is an ideal way of explaining the missing mass. Superstring theories were developed in the pursuit of "Grand Unified Theories" and "Theories of Everything." The evidence for dark matter (from astrophysicists) came about independently of these theories. The close correlations between shadow and dark matter were noticed only later. In fact, dark matter is often also described as "shadow" and "mirror" matter. To some extent, this cross-validates both superstring and dark matter particle theories.

E6xE6 Superstring Theory

While the E8xE8 superstring theory makes some interesting observations about shadow universes, it has been realized more recently that there are some technical limitations to it. The main criticism is that the symmetry group relating to this theory does not have any representations in complex vector space or chiral representations. This means it would not be possible to describe fully the physics in the ordinary matter universe. However, a subgroup of this theory, the E6xE6, is the only one among five exceptional Lie groups that has these representations. One E6 component of this subgroup will be able to explain all the particles and forces in our ordinary matter universe, while the other component would represent all the forces and particles in a shadow universe – as described by physicists, discussed above.

Mirror Universes in Science

Some interactions of elementary particles seem to be biased to the left – as if they only interacted with an object but not its mirror image, for e.g., in beta decay. Beta decay is a type of radioactive decay in which a proton is transformed into a neutron, or vice versa, inside the nucleus of an atom, by the weak nuclear force. The weak force is one of the four fundamental forces in the standard model of particle physics. Since, beta decay is biased to the left, it may be interpreted as a violation of symmetry by the weak force. If we took a step back, however, we can ask the question: if our universe is a left-handed universe, is there a "right-handed mirror universe" that restores the symmetry from a larger multiverse perspective? In other words, could there be a mirror universe whose symmetry is the reverse of our universe, and is biased to the right?

Z K Silagadze and Robert Foot think so. In his paper "Mirror objects in the Solar System?" Silagadze argues nobody doubts that these violations in symmetry are firmly established experimentally. But this does not necessarily mean that nature is asymmetric. Symmetry can be restored if we can imagine a mirror world where for every ordinary particle there exists a "mirror" particle. Foot agrees and he says that the left-handedness of ordinary particles can be balanced by the right-handedness of mirror particles, in his book "Shadowlands." This left or right reversal in the mirror universe is equivalent to a spatial inversion (imagine what happens to the appearance of your right hand when reflected in a mirror.) As discussed previously, it is referred to as "parity."

One solution to the observed asymmetries within our own universe is therefore to postulate that there is a mirror universe which restores overall symmetry. No doubt, this idea of a mirror universe ties in quite nicely to the idea of a shadow E8 or E6 universe, discussed earlier. One E6 component is our universe, and the other E6 component is this mirror universe that restores overall symmetry. Silagadze equates mirror matter with shadow matter. It is easy to see why: mirror matter is ordinarily invisible and interacts with ordinary matter mainly through the gravitational field, just like shadow or dark matter. A range of persuasive evidence exist for mirror matter and many scientific papers have been written about it. Scientists take mirror universes and particles seriously and are conducting scientific experiments to detect it, for e.g., in Oak Ridge National Laboratory in Tennessee.

Shadow and Mirror Universes in the Metaphysical Literature

Shadow Universes in the Metaphysical Literature

The theories relating to dark, shadow or mirror matter and their universes bear a strong resemblance to the theories and direct observations by metaphysicists and mystics. Bernard Carr is a professor of mathematics and astronomy at Queen Mary University of London. He points out in a speech, given at the Euro-PA Conference in November 2003, entitled "Is there Space for Psi in Modern Physics," the importance of considering the metaphysical evidence:

> The higher dimensional "reality structure" required to accommodate psychic experiences is intimately connected with the higher dimensional space invoked by modern physics....This "universal structure"...necessarily incorporates physical space but it also includes non-physical realms which can only be accessed by mind ... This is very far removed from the naive view of reality adopted by ... reductionist materialists who reject psychic phenomena out of hand.
>
> Bernard Carr, Mathematical Physicist,
> Queen Mary University

This book takes science-based metaphysical evidence seriously and shows that scientists can benefit by reflecting on these in constructing their theories. If shadow, mirror, and dark universes are taken to mean universes which share the same spacetime landscape as ours, then we have to include the "super-physical universes" popularized by metaphysicists. These universes are also the "dark sectors" that scientists currently classify as part of our own universe. Since each dark sector contains different particles and largely different fundamental forces, and potentially different spacetime signatures, it is more logical to describe these dark sectors as partially interacting "universes." From this perspective, our current measurable universe would be considered a multiverse. In this "local multiverse," the low-dimensional ordinary matter 3d universe, and its dark etheric double (i.e., its E6 mirror component,) sit inside the valleys of a higher-dimensional landscape which interpenetrates it and looks dark (i.e., invisible) to us. Super-physical universes can be imagined extending vertically up to higher energy levels, with an increasing number of spatial dimensions. These ideas are not new to metaphysicists, as summarized by the commentary below:

> The physical world is only a small part of the entire spectrum of matter. It is the densest, and most concrete of a series of worlds ranging from the extremely tenuous "super-physical" to the solid physical. This is an idea found in ancient Egyptian mysteries, Hinduism, and Buddhism.
>
> Shirley Nicholson, 1977

Metaphysicists believe that there is not only one other "shadow" universe, but a number of interpenetrating shadow universes. Furthermore, the "E6" shadow universe that coexists with "our" universe as its double, within a single spacetime of three space and one time dimensions, had already been identified by metaphysicists as the duplicate mirror "physical-etheric" universe. Many cultures, including the ancient Egyptians, already knew about this universe through direct observations, almost five thousand years ago. They identified a "spiritual double" (called the "ka") accompanying the ordinary matter body, also known as the physical-etheric double. This double lived in a parallel duplicate universe. This mirror universe is discussed further, below, and in chapter 14.

> Every solid, liquid, and gaseous particle of the physical body is surrounded with an etheric envelope: hence the etheric double, as its name implies, is a perfect duplicate of the dense form.
>
> Annie Besant, 1896, Metaphysicist

In 1893, metaphysicist Besant, who presumably had first-hand knowledge of other universes, described universes, other than the physical, peopled with intelligent beings. It is crowded, like our world, with many different types of life-forms, as diverse as a blade of grass from a tiger, or a tiger from a man. "It interpenetrates our own world and is interpenetrated by it, but, as the states of matter [i.e., the elementary particles and fundamental forces] of the two worlds differ, they co-exist without the knowledge of the intelligent beings in either." This is almost identical to the descriptions given by scientists currently, including the ones quoted above.

When symmetry broke at the big bang, one of the E6 components became our universe, and the other became a slightly higher energy mirror universe that metaphysicists are already familiar with as the physical-etheric universe. But there are even higher energy shadow universes and their mirror counterparts, beyond the physical-etheric universe. The next in the energy spectrum is the astral universe. Leadbeater shares his observation about the shadow astral universe:

> Astral matter, being much finer than physical matter, interpenetrates it. Thus, a being living in the astral world might be occupying the same space as a being living in the physical world; yet each would be entirely unconscious of the other and would in no way impede the free movement of the other. The different realms of nature are not separated in space but exist about us here and now. Each world has its inhabitants, whose senses are normally capable of responding to the undulations of their own world only.
>
> Charles Leadbeater, 1912, Metaphysicist

This echoes the current descriptions given by mainstream physicists of shadow matter and interpenetrating shadow universes under superstring and other modern theories of the multiverse – including eternal inflation theory and M-theory (which will be discussed in chapter 11.)

Mirror Universes in Metaphysical Literature

In his book, "Journeys Out of the Body," Robert Monroe, a radio broadcaster who was qualified with a degree in engineering, explains (below) that when he was out of his ordinary matter body, he tried to touch his head, but he felt his foot instead (top-down reversal.) When he tried to touch his left toe, he felt his right toe (left-right reversal.)

> After lifting out, I carefully examined the physical body on the bed. I reached down to touch my physical head, and my hands touched feet! I felt my toes. My left big toe has a thick nail — this left toe did not! The big toe on the right foot did have the thick nail. Everything was reversed, like a mirror image. There is a possibility that the Second Body is a direct reversal of the physical.
>
> Robert Monroe, Experimental
> Metaphysicist and Engineer

The reversal of parity occurs because the lowest energy dark universe (or the "physical-etheric universe" in metaphysical parlance) appears to be a right-handed mirror universe. It is both left-right reversed, as well as top-down reversed. It is interesting that a flat mirror may be perceived by our brains to reverse left and right. However, a vertically curved concave mirror not only reverses left and right, but also turns the image upside down at a certain distance. Tom Siegfried, a science journalist, says that a mirror universe could exist in the same space as our universe if the "mirror" is sufficiently distorted, sort of like in a carnival fun house, where curved mirrors are used to make amusing distorted images. It looks like the local multiverse could probably be very much like a carnival fun house, whose purposes include education and entertainment.

But how could Monroe tactically sense the ordinary matter body, while having a visual representation which was spatially reversed from the physical-etheric body? As already noted in the author's book, "Our Invisible Bodies," dark plasma becomes elongated into the shape of a "cable" while it is being stretched by the detached physical-etheric body at it moves away from the physical-etheric envelope around the ordinary matter body. The tactile signals from the envelope will therefore be transmitted to the detached physical-etheric body. Since the envelope is strongly coupled to the ordinary matter body, the tactile signals will not be reversed, while the visual signals from the largely detached physical-etheric body would be. This betrays multiple (in this case, double) overlapping streams of consciousness.

In an experiment conducted by William Tiller, Professor Emeritus of Stanford University, this parity reversal has also been observed. Tiller, reporting that during the experiment the children saw the spectrum with their physical eyesight *and* also a bent-up spectrum with their "auric eyesight." This evidences again multiple (in this case, double) overlapping streams of consciousness. He concludes that this experiment confirms "a type of 'mirror' relationship functioning between the physical level and the next, more subtle [i.e., physical-etheric level] of reality." He also reports that Qigong practitioners appear to be projecting a type of magnetic energy with inverse characteristics to a regular magnetic field. If so, they could be projecting what we would now call dark magnetic fields, which are mirror images of ordinary magnetic fields.

There was also a foreground-to-background reversal that was observed by metaphysicist Barbara Brennan (formerly, a physicist at NASA,) when observing dark plasma bodies. These various spatial reversals (in the x, y, z, and perhaps other axes in higher-dimensional space,) discussed above, can make analysis of events in higher energy universes very tricky. Even Alice's Wonderland would seem tame compared to these universes! (Incidentally, another term used in science to refer to shadow and mirror matter is "Alice matter.")

Robert Monroe says that when someone is undergoing a dramatic shift in consciousness, such as having an out-of-body experience, the body's polarity, or electromagnetic field, reverses itself. In 1939, W E Burge of the University of Illinois found that the voltage measured between the head and other parts of the body became more negative during physical activity, declined in sleep, and reversed to positive under general anesthesia.

Robert Becker MD, a towering figure in bioelectromagnetics, and his collaborator, Charlie Howard, also found that the back to front current in the human head varied with consciousness. It was strongest during heightened physical or mental activity, declined during rest, and reversed direction in both normal sleep and under anesthesia. These polarity reversals seem to betray shifts in the locus of consciousness from the body to its reversed mirror body. In normal waking consciousness, physical awareness is located in the ordinary matter body. During deep sleep, this awareness shifts to the etheric double (inhabiting a reversed universe.) The physical-etheric body lives in a mirror universe, and is not only reversed in parity, but also polarity or charge.

Matter dominates this universe at cosmic scales because anti-matter particles are sequestered in the higher energy physical-etheric mirror universe. These parity-reversed anti-matter particles will interact only weakly with ordinary matter particles – making them a component of dark matter. Detection techniques for dark matter particles must therefore be more sophisticated to take this into account or else they would escape detection.

CHAPTER 7

Our Plasma Universe

Plasma Everywhere!

Although rare in our close environment, more than 99 per cent of our ordinary matter universe is in the plasma state! The Sun and other stars, as well as the invisible gas between them, are all in this state. A negligible amount of matter makes up the cold celestial bodies such as the Earth and other planets, the moons, asteroids, comets, meteors, and dust grains in space. Space probes have found plasma in the shells around Earth (including the ionosphere and plasmasphere) — and also farther out — blowing out from the Sun and guided by weak magnetic fields in interplanetary space. In other words, atomic and biomolecular matter, which is so abundant on this planet, and which makes up the bodies of a myriad of life-forms, including our own ordinary matter body, is not representative of the state of matter in the ordinary matter universe. In fact, it gives us a very misleading picture of our universe.

What is Plasma?

To understand what plasma is, think of a hydrogen atom. As a whole, it is neutral. It consists of a positively-charged proton in the nucleus with a neutralizing negatively-charged electron forming a cloud around the nucleus. If enough energy is supplied, the electrons will be able to break free from their atoms, overcoming their binding energies. They then separate to form a soup of negatively-charged electrons and positively-charged protons. This soup is overall neutral and exhibits collective behavior because of the electric and magnetic fields. In other words, the particles are the same as in atoms, only the configuration has changed. This plasma-like collective behavior can be observed even when only 1 per cent of atoms in a given volume lose their electrons, i.e., when they become "ionized."

Since there are large soups of free electrons in plasma, it is an excellent conductor of electricity, far exceeding the conducting properties of metals such as copper or gold. If the light electrons and heavy (positive) ions are separated (for e.g., due to gravity or mechanical forces,) they produce electric fields. This may generate currents of moving charged particles, which give rise to magnetic fields, as well as electromagnetic radiation. The magnetic fields guide and confine the particles, while the electric fields accelerate them to high speeds. The charged particles in plasma react strongly to electromagnetic forces, generating a complexity in structure that far exceeds that found in solids, liquids, or gases.

Some examples of everyday manufactured objects in the plasma state include fluorescent lamps and neon lights. In nature, there are many examples. On Earth, the aurora borealis (or Northern Lights) and the plasma that is produced by lightning are good examples. As noted in chapter 3, Earth's ionosphere and plasmasphere contain both hot and cold plasma. The Sun and other stars are balls of plasma. The solar wind is a plasma of charged particles.

Complex Plasma

The plasma in space not only includes charged sub-atomic particles, but also much larger dust particles, which have become negatively charged due to the accumulation of lighter electrons attached to it. This type of plasma is called "dusty plasma," or "complex plasma." Due to the presence of the charged dust, complex nonlinear dynamics occur. These result in intricate processes and interesting features in complex plasma, including structures that look like DNA. Complex plasma has also the characteristics of a liquid-crystal.

Plasma and Magnetic Fields

We know from basic electromagnetics that currents (i.e., moving charged particles) generate magnetic fields around them. The natural tendency of plasma to carry currents is therefore an important source of magnetic fields. For e.g., the solar wind is a plasma which consist of moving charged particles that generates a weak magnetic field. Since plasma is pervasive throughout the universe and may give rise to currents, scientists believe that virtually all ordinary matter in the universe (in galaxies and galaxy clusters) is magnetised. But magnetic fields are also found outside galaxies and galaxy clusters where there is hardly any ordinary matter. Where did these come from?

As discussed in chapter 4, the cosmic dark matter web provides the gravitational scaffolding for invisible hot ionized gas to flow along its filaments, giving rise to hot inter-galactic filaments of ordinary plasma. These hold quasi-neutral currents of oppositely-charged particles that generate ordinary inter-galactic magnetic fields, and also magnetic fields between galaxy clusters. Additionally, if the cosmic dark matter web were in the plasma state and projected a weak dark electromagnetic force, it could generate weak dark magnetic fields, even between galaxies and galaxy clusters. If we do detect these dark magnetic fields, for e.g., through the presence of dark photons that convert to ordinary photons, it will provide evidence that dark matter (in the vicinity of ordinary matter) is in the state of a dark plasma.

While currents of ionized matter give rise to magnetic fields, the fields also influence, and structure ionized matter. They guide charged particle movements, which sweep up both charged and neutral matter, and shape them. Scientists studying patterns in plasma in the laboratory and in space infer that there are complex nested hierarchies of magnetic fields in the universe, mirrored by an associated hierarchy of electric currents. Magnetic fields and electric currents "feed" on each other and breed new fields and currents. From the Earth's plasmasphere, out to the most distant intergalactic regions, all cosmic plasma are penetrated by magnetic fields that influence their physical properties in various ways to produce signature features associated with plasma. These are discussed below.

Signature Features of Plasma

Filamentary Currents

For plasmas immersed in strong magnetic fields, currents of charged particles tend to flow around the magnetic field lines, which act like wires guiding them. These types of currents are called "Birkeland currents." They occur both on Earth and in space.

On some clear nights, in the Arctic and Antarctic regions, the sky is filled with undulating sheets of luminescent colors that move and dance called "auroras." It is the visible manifestation of huge, invisible Birkeland currents plunging into Earth's atmosphere — guided by Earth's magnetic field lines. On a much larger scale, in space, one of the most compelling pieces of evidence for the existence of Birkeland currents, between clusters of galaxies, came from the discovery of faint super cluster scale radio emissions at 326 MHz between the Coma and Leo clusters of galaxies. Electrons and ions are accelerated in opposite directions, giving rise to an overall neutral current along the magnetic field lines.

The magnetic fields move with the plasma and their strength increases as the plasma becomes more dense. Any rotation of the plasma would twist the embedded magnetic field causing helical Birkeland currents to appear when charged particles rush through, guided by the field lines.

Vortexes

The existence of coherent vortical structures is a characteristic feature in magnetized complex plasma that are found in space and seen in laboratory experiments. These vortexes (or vortices) arise when filamentary Birkeland currents cross and pinch to form nodes, which give rise to intense magnetic fields. The currents may also pinch when the magnetic force pulls them together, when they are alongside each other The angular momentum from charged particles, moving in different directions in the filaments, and the magnetic torsion that results, gives rise to rotating vortexes.

Double-Layers

To ensure that the plasma is overall neutral, a double layer forms in plasma around any significant excess of positive or negative charge, which may give rise to currents. It is a region consisting of two oppositely charged parallel layers. This generates a voltage drop and an electric field across the layer, which accelerates electrons and positive ions in opposite directions. This gives rise to currents in opposite directions but overall neutral, and electromagnetic radiation of a specific frequency. Double layers are found in plasmas in the laboratory, as well in Earth's plasma environment (for e.g., in the ionosphere and the plasmasphere,) and around stars, including the Sun.

Is Self-Interacting Dark Matter in the State of a Plasma?

Self-interacting (i.e., charged) dark matter, which makes up about 15 per cent of all matter in the universe/multiverse, was discussed in chapter 5. According to metaphysicists, the electromagnetism, associated with invisible subtle matter (which is believed to be dark matter by the author,) is weaker than what we experience in the ordinary matter universe. In our current context, to distinguish it from ordinary electromagnetism, we will call it "dark electromagnetism."

The dark electromagnetic force would be weaker and shorter-ranged in higher dimensional universes because it will have to spread out over more than three spatial dimensions. This means it will propagate, not based on the inverse-square law, but the inverse-cube, inverse-quartic, or inverse quintic law and higher, and therefore drop off faster at shorter distances. Furthermore, astrophysical observations show that dark matter particles move much slower than the speed of light in the ordinary matter universe. We will therefore assume that self-interacting (charged) dark matter particles move slower than the speed of light (as measured in the ordinary matter universe,) and project a weak dark electromagnetic force that is shorter ranged (compared to ordinary electromagnetism.) It will become progressively weaker as we ascend to higher-dimensional universes.

Since the dark electromagnetic force is weaker and shorter-ranged, these dark matter particles, although charged, will not interact if the particle density is very low, for e.g., outside galaxies, and generally in the regions of the galaxy's dark halos which are far away from ordinary matter. Here, the inter-particle distance will be very large. This means the probability of colliding or interacting with other slow-moving particles will be extremely low. It would be difficult for the particles to feel the weak dark electromagnetic force at large distances. Furthermore, as these particles live in higher energy universes, they will have more kinetic energy to separate from other particles, which would lower the particle density.

However, when a high density of ordinary matter gravitationally attracts dark matter particles, the inter-particle distance will be significantly reduced, and the matter will condense to a degree where the weak electromagnetic force can operate. This higher density soup of positively and negatively charged dark matter particles will effectively be a "dark plasma," i.e., a plasma of charged dark matter particles. The higher density will lead to collisions between dark and ordinary matter particles in the regions of the galaxy where ordinary matter is present in abundance. In spiral galaxies, like the Milky Way, this would in the central region of the dark matter halo.

The collisions will make the dark matter particles lose kinetic energy, cool down, and become gravitationally bound to ordinary matter. Consequently, the dark matter will clump even more. (Although the word "cool" is used here, it can be very hot relative to room temperature. The temperature referred to in scientific literature is measured using the estimated speed of particles, relative to the speed of light, and not to room temperature. We cannot feel this temperature directly in the ordinary matter universe as dark matter does not radiate light or heat within the ordinary electromagnetic spectrum.)

The bulk of the self-interacting dark matter particles will not combine to form atoms. This is because the weak electromagnetic force and the higher kinetic energy of the particles (as discussed above) will make it difficult for dark protons to rein in dark electrons. Additionally, if the particles had a "strong force," like ordinary matter, this would be weaker. This is because under supersymmetry theory, the strong force dramatically weakens at higher energies. This means dark protons and dark neutrons cannot come together to form complex or heavy atoms in a nucleus in higher-energy universes (beyond the 3d universe.) In the lowest energy dark universe (i.e., the 3d physical-etheric universe) there is a possibility of forming simple atoms. There would also not be any significant annihilations with anti-dark matter. Anti-dark matter would not be present at cosmic scales (just as in the ordinary matter universe) as it would be sequestered in a mirror universe. This leaves us with what would be known technically as "asymmetric dark matter."

This gives us a basis to consider most of self-interacting dark matter to be in the state of a dark complex plasma. Dark complex plasma can be modeled by using what we know about ordinary complex plasma.

Further evidence that supports this conclusion is that signature features of plasma have been seen in dark matter. Dark matter halos constructed for elliptical galaxies, using gravitational lensing, reveal the presence of faint concentric shells. Concentric shells are signature features of plasma crystals (which are discussed in more detail in the next chapter.) Mach cones, which arise in plasma vortexes, have also been seen in computer simulations of dark matter density and distribution.

Additionally, signature features of plasma have been recorded in the metaphysical literature. The literature is rife with references to "subtle bodies," that accompany the ordinary matter body, having chakras and nadis (or acupuncture meridians.) These are unmistakable features associated with plasma – chakras being plasma vortexes (or vortices) and nadis being filamentary currents in plasma (more specifically Birkeland currents.) Subtle bodies, described by metaphysicists, display coronas, project beams of light and can generate heat and light – all features of plasma bodies (such as the Sun and rotating neutron stars – also known as pulsars.) The "auric sheaths" observed by experimental metaphysicists around the ovoids of dark plasma bodies can be easily identified with the "double-layers" in plasma. They serve the same purpose.

In 1910, experimental metaphysicist Leadbeater said that on higher planes (i.e., the dark sectors,) everything is what down here we would call luminous (plasma is luminous through the spontaneous emission of photons,) and above a certain level everything may be said to be permeated by fire. "Try to think of a fire which does not burn, but is in a liquid form, something like water," he says. He adds, "all astral matter is in itself luminous, though an astral body is a sphere of living fire." A *liquid fire*, or a *cold fire* (which some have used to describe a "nonthermal plasma" today) is a perfect description of plasma. Under the study of magnetohydrodynamics in plasma physics, plasma is treated as a fluid. When Leadbeater was describing this state, the word "plasma" was not yet invented by Nobel Laureate in Chemistry, Irving Langmuir, in 1929. Hence, he had some difficulty in pinning down a word for it so elaborated in terms of more familiar terms like "fire." (Incidentally, fire is considered a "near-plasma.")

In 1914, Leadbeater classified matter as follows: "In the matter of the physical world the seven subdivisions are represented by seven degrees of density of matter, to which, beginning from below, we give the names solid liquid, gaseous, *etheric, super-etheric, subatomic and atomic*." Today, we call the state of matter that comes after gases, plasma. This implies that *etheric* is dense dark plasma, *super-etheric* medium-dense dark plasma, *subatomic and atomic* relate to low density, diffuse plasma where individual particles are far apart from other particles. All this indicates that self-interacting dark matter (just like ordinary matter) is largely in the state of a plasma.

Photonics

At higher energies, as plasma becomes more radiant, photonics, or the study of light particles, and associated technologies, become important. These technologies relate to lasers, optical fibers, lenses, optical sensors, photonic propulsion, quantum computing, and data storage, among others. A subset of photonics is plasmonics, which takes advantage of light's relationship with plasma.

Conclusion

We learnt that our *low energy* ordinary matter universe is more than 99 per cent composed of plasma. If that is so, then it is highly probable that the even *higher energy* dark universes would be composed of self-interacting dark matter largely in the form of plasma, with an even higher rate of ionization. Since, these dark plasma universes interpenetrate our ordinary matter universe, would there be dark matter counterparts of ordinary matter planets, i.e., shadow planets composed of dark plasma, gravitationally bound to ordinary matter planets? The evidence indicates that there is a high probability that this could be so. We can call these dark "plasmaspheres." We learnt at the beginning of this book that heavens and hells in religion and in history are often structured into many different levels – as concentric shells around Earth. Do dark plasmaspheres also structure themselves into many levels? If so, we will be closer to supporting the claim that the popular cultural heavens and hells do indeed interpenetrate the ordinary matter Earth. This is discussed in the next chapter.

CHAPTER 8

❋

Dark Plasmaspheres

Shape of Dark Plasmaspheres

We noted that Earth's ordinary matter magnetosphere and plasmasphere takes on a "tear-drop" shape in chapter 3 — in other words, the side facing the Sun is blunt and the opposite side has a long tail. This is due to the pressure from the Sun's solar wind. Furthermore, it expands and contracts in different conditions. Earth's counterpart dark plasmaspheres would similarly take on a tear-drop shape which expands and contracts according to space weather in the dark sector, including those caused by dark plasma winds from the dark Sun and the dark galactic halo (particularly in the vicinity of the ordinary matter galactic disk, where the density of self-interacting dark matter particles is greatest.) Additionally, like our visible oceans, these invisible oceans of dark plasma would be subject to tides because of the gravitational interactions within the Earth-Moon-and-Sun system. The dark plasmaspheres would therefore be constantly size and shape-shifting.

The dark geomagnetic field, which is embedded in the dark plasmasphere, moves with it, changing its size and shape in tandem. The field is generated by the plasmasphere as it rotates. Just like Jupiter's gas envelope, plasmaspheres do not have a hard surface. Plasma life-forms will move about more like fish in the oceans or birds in the air, using ocean and air currents, respectively. Each of Earth's dark plasmaspheres, or "Dark Earths," can be considered a planet, a "protostar" which shares the same gravitational field with the visible Earth and co-rotates with it. The dark plasmasphere is a ball of plasma – just like the Sun. However, it does not generate nuclear fusion. Hence, it could be considered a permanent protostar.

Just like dark plasma bodies, the plasmasphere can interact with the frequency of the ambient dark light, resulting in the plasmasphere having different appearances to an observer. If its internal composite plasma frequency is higher than the ambient dark light, it will shine like a metal ball, if it is lower it will become transparent and disappear, and if it's the same, it will be a black ball that will disappear against the darkness of space. Since, the plasma frequency is directly related to its density, and its density keeps changing due to space weather and gravitational interactions, the Dark Earths will oscillate in brightness, becoming visible and then invisible alternately – causing a scintillation.

We expect a generically similar configuration and shape for dark plasmaspheres that interpenetrate other ordinary matter planets in the Solar System, as well as exoplanets in the Milky Way and elsewhere.

Density Profile of Dark Plasmaspheres

> Astral matter gravitates towards the center of the Earth, just as physical matter does.
>
> Charles Leadbeater, Experimental Metaphysicist, 1910

In 1910 Leadbeater reported that "the densest aggregation of astral [or dark] matter is within the periphery of the physical body of a man." Similarly, in the case of Earth's dark astral plasmasphere, he reports that the greater part of its astral (or dark) matter is gathered together within the limits of its visible ordinary matter sphere. (If we extrapolated this to the Milky Way, the greatest density of dark matter would be within the ordinary matter disk. This would produce a dark disk inside the ordinary matter disk, consistent with the theory proposed by theoretical physicist Lisa Randall of Harvard University.)

The "Lambda Cold Dark Matter (or Lambda CDM)" model supports the density profile observed by Leadbeater. This popular scientific model explains very well many large-scale structures in the universe and predicts that dark matter rapidly increases in density towards the center of a galaxy. Scientists use it to study the distribution and density of dark matter and it is supported by many observations. In a 2003 study using the Sloan Digital Sky Survey, astrophysicists modeling the motion of dark matter within the galaxy, say that each dark matter clump in the galaxy had a density that peaked in the center and fell off toward the edges. Observations with the

Chandra X-ray Observatory also support the model for galaxy clusters. The data obtained by John Arabadjis and Mark Bautz of the Massachusetts Institute of Technology (MIT), along with Gordon Garmire of Pennsylvania State University in State College, found that the density of dark matter is greater the closer it is to the center of the galaxy cluster.

Nevertheless, other studies indicate that there is some deviation from the Lambda CDM model at smaller sub-galactic scales. These show that density in the centers of galaxies is not as high as predicted by the model. In these cases, it has been found that there is an additional factor to consider i.e., there seems to be observed density correlations between ordinary and dark matter. These correlations suggest that dark matter is interacting with ordinary matter in some way. A new fundamental fifth force has been conjectured to explain this. Based on this correlation, we would expect, with all other factors being constant, a higher density of dark matter in the centers of galaxies where there is a higher density of ordinary matter (in line with the cold dark matter model.) However, the converse would happen where there was very little ordinary matter. This realization does not invalidate the cold dark matter model. It fine-tunes our understanding of how the model should be applied at smaller scales, after considering gravitational interactions with ordinary matter.

For non-astronomical small bodies, such as the human body, the cold dark matter model would not be directly applicable. In this case, we would need to consider an attractive force between dark and ordinary matter. Based on this, we would expect a dark plasma body, associated with a human, to consist of a dense humanoid figure inside a low-density ovoid bubble or halo. (Similar to a dark disk inside the halo of our galaxy.) The density of the dark plasma humanoid figure would correlate to the density of the dense ordinary matter body. Similarly, with respect to the ordinary matter Earth, we would expect most of the dark matter to be within the crust, as observed and reported by metaphysicist Leadbeater in 1910. Hence, dark matter would be densest within Earth's core and mantle, where the densest ordinary matter exists.

If we compare equal volumes of dark and ordinary matter, the mass of the dark matter would be much less. However, in large scale structures, for e.g., a spiral galaxy, we typically find a small volume of ordinary matter sitting inside a huge volume of dark matter. In this case, the total mass of the dark matter would be many multiples of the ordinary matter, and the ordinary matter would be subject to the gravitational mould imposed by dark matter. However, in stellar systems and planets composed of ordinary matter, which have dark matter in their vicinity, the gravity from the higher mass of the much denser ordinary matter dominates. While dark matter tells ordinary matter what to do at large scales, ordinary matter tells dark matter what to do at small scales.

Density of Dark Plasma Correlates with Density of Ordinary Matter

Experimental metaphysicist, Leadbeater, notes, "each subdivision of physical matter has a strong attraction for astral matter of the corresponding subdivision." In other words, an ordinary matter solid will attract much denser dark astral plasma than an ordinary matter liquid. He explains: "If I have a glass of water standing upon a table, the glass, and the table, being of physical matter in the solid state, are interpenetrated by astral matter of the lowest subdivision [i.e., an astral solid.] The water in the glass, being liquid, is interpenetrated by astral matter of the sixth subdivision [i.e., an astral liquid,] while the air surrounding both, being physical matter in the gaseous condition, is entirely interpenetrated by astral gaseous matter."

We may use relative terms to distinguish low and high density dark plasma, by calling it a "plasma-liquid" or a "plasma-solid," respectively. However, a dark plasma solid is extremely low in density and even more tenuous than even ordinary matter gases. Leadbeater cautions, "...even the astral solid is less dense than the finest of the physical ethers." The important point is that the density of dark plasma correlates with the states of matter of ordinary matter. An ordinary matter solid will attract a dark plasma-solid, an ordinary matter liquid will attract a dark plasma-liquid, and an ordinary matter gas will attract a dark plasma-gas. This correlation has also been observed in astrophysical observations of galaxies. It suggests that there is an attractive fundamental force between ordinary matter and dark matter particles, which has not yet been detected by science. Hence, we would expect a correlation in the density between the shells of dark plasmaspheres and the shells of the ordinary matter Earth, discussed in chapter 3. We would expect highly dense dark plasma in the cores of dark plasmaspheres, which would correlate with the high-density core of the ordinary matter Earth.

Some physicists have already theorized that there could be a large concentration of dark matter in the Earth's core due to Earth's gravitational attraction. Physicist David Peat says the best calculations suggest that Earth's core could contain as much as 10 percent shadow matter. Metaphysicist Leadbeater had observed that the "fiery" kundalini, a dense form of super-physical plasma, originates from the center of the Earth — distinguishing it from "prana" or "qi" which he observes emanates from the Sun (as discussed in chapter 4.) He says, "The force of kundalini in our bodies comes from deep down in the Earth. It belongs to that terrific glowing fire of the underworld. That fire is in striking contrast to the fire of vitality which comes from the Sun." The higher density super-physical matter, consisting of heavier and denser dark plasma, gravitates and sediments below the Earth's crust.

Formation of Concentric Shells in Dark Plasmaspheres

The formation of a dense core of heavy dark matter particles facilitates the formation of double-layers and concentric shells in the dark plasmasphere. To understand how this will happen, we review an experiment with plasma crystals by H. Thomas, conducted in the laboratory. These crystals were in the form of assemblies of particles which were held in a crystal-like array by a plasma of weakly ionised gas. When the assembly of microscopic particles was contained between two electrodes and illuminated by a laser beam, it could be seen, even with the naked eye, that the particles naturally arranged themselves regularly into as many as eighteen *planes* parallel to the electrodes. In another more recent experiment, the particles in a plasma crystal arranged themselves into neat *concentric shells*, to a total ball diameter of several millimeters. These orderly Coulomb balls, consisting of aligned, concentric shells of dust particles, survived for long periods.

One of the most important properties of any electrical plasma is its ability to "self-organize" — that is, to electrically isolate one section of itself from another, based on their properties. The isolating wall is called a "double-layer." For e.g., if the voltage difference from one electrode to the other becomes large enough in a plasma, a thin double-layer will form somewhere in the middle of the tube. The plasma on one side of the double-layer (the side toward the anode) will have approximately the same voltage as the anode. The plasma on the cathode side of the double-layer will have essentially the same voltage as the cathode. The two halves of the plasma are then electrically isolated from one another by the double-layer. No electrostatic force is felt by particles on either side of the double-layer due to charges on the other side of the double-layer. The strongest electric fields in the plasma will be found within the double-layer.

As discussed in the previous section, each plasmasphere would be expected to have a dense center. Due to gravitational forces, we would expect the heavier (say, positively-charged) self-interacting dark matter particles to settle and be compactified within the center of the Earth, as a quasi-spherical body. The lighter (say, negatively-charged) particles would surround the plasmasphere (i.e., the Dark Earth) as a dark atmosphere. (This separation of charges would not happen in the ordinary matter Earth as oppositely charged particles would be bound within atoms and molecules.) In between the two regions (i.e., the dark core and the surrounding dark atmosphere) there would be a mixture of both types of particles.

The two regions would therefore represent natural electrodes, with the quasi-spherical body in the center of the Earth being the anode. When the voltage difference between the two electrodes becomes very large, it would generate a series of quasi-spherical double-layers within the dark plasmasphere, with electric fields within. The strength of the electric field in each succeeding double-layer would become weaker towards the middle, i.e., between the regions that represent the electrodes. The double-layers would separate plasma with different properties and effectively create *concentric shells* of plasma with different densities, temperatures, and frequencies. The number of shells will depend on the voltage potential of the electrodes – the higher the potential, the greater the number of shells. Since the voltage potential can change over time, the number of shells would also vary. The plasma between the double-layers (i.e., within the shells) would be overall neutral.

One of the unique characteristics of space plasma, revealed by satellites and space probes, is its tendency to separate and form sharp boundaries between plasmas with different properties, in the form of electrified "sheaths" (i.e., double-layers) around concentric shells. These concentric shells can be seen in space plasma. The Cat's eye Nebula (NGC 6543) is a nebula in the constellation of Draco. Images, taken with Hubble's advanced camera for surveys, reveal the full beauty of a bull's eye pattern of eleven or even more concentric rings, or shells, around the Cat's eye. Romano Corradi (from the Isaac Newton Group of Telescopes, Spain) and collaborators, in a paper published in the European journal of astronomy and astrophysics in April 2004, have shown that the formation of these shells is likely to be the rule rather than the exception. Concentric shells can also be seen in dark matter halos in galaxies (this has already been discussed in chapter 7.)

In chapter 3, we learnt that historically, in religion, many different levels of heavens and hells were discussed. Each human-linked heaven at a specific level would represent an ecological niche within a specific shell in Earth's dark plasmasphere which interpenetrates the ordinary matter Earth. Almost every metaphysicist agrees that there are many different super-physical planes i.e., every dark plasmasphere would be divided into different shells with different properties. This division into concentric shells, which represent different levels, is a natural consequence of the behavior of magnetized plasma.

Since the dark plasmasphere changes size and shape in response to space weather and gravitational interactions, the concentric shells will also deform in tandem with the plasmasphere and the magnetic field lines embedded in it. As reported by metaphysicists, the shells are composed of decreasing densities of dark matter, moving out from the center of the Earth. As we move-up the energy ladder, each shell becomes less dense, more tenuous, and higher in energy, frequency, and ionization rate. Joel Whitton and Joe Fisher, noted in their book "Life between Life," that "each higher plane is lighter and brighter than the one before."

Dark Plasma Domains and Electric Fences

The double layers, separating the concentric shells in the plasmasphere, will not prevent Birkeland currents from penetrating it if the energy of the particles in the currents within the double layer is less than that of the Birkeland currents. (Just as highly energetic charged particles in Birkeland currents plunge through the double layers of the ordinary matter Earth's plasmasphere to generate auroras in the ionosphere.) The fine filamentary dark magnetic field lines of the dark plasmasphere could also connect with those of the individual's dark plasma body to deliver charged dark matter particles, which supply energy to the dark plasma body.

Where adjacent plasmas have different properties, double layers will form to separate them into smaller cells within the shells. The cellular structure of plasma can be clearly seen in the granulation of the Sun. These cells would have different frequencies due to the different properties. The Hindu mystic, Paramahansa Yogananda, explains that various "*spheric* mansions" and "*vibratory* quarters" are provided to astral beings. These are cells vibrating at specific frequencies, which can be quite large. This is reminiscent of the saying attributed to the historical Jesus: "In my father's house are many mansions [i.e., large dwelling places.]" The double layers or sheaths, enclosing the shells, function as electric fences — making it difficult for plasma bodies to wander off to another shell. However, Yogananda clarifies, "Advanced beings are, however, able to cross *boundaries*" [emphasis added.] These boundaries, or double layers, would have currents of charged particles at higher energies than the regular inhabitants. Hence, they would be opaque and impenetrable to these inhabitants. However, they would be transparent to higher energy bodies. Although dark plasmaspheres interpenetrate, the shells and cells within each plasmasphere generally do not.

Interactions between Plasmaspheres

General weather conditions in the ordinary matter Earth does not affect other interpenetrating higher-dimensional spheres. This is consistent with scientific theory (to be discussed in chapter 11) as the particles and forces relating to the different universes, in which these plasmaspheres exist, are confined to the relevant universe. Metaphysicist Leadbeater notes, "Atmospheric and climactic conditions [in the ordinary matter visible Earth] make practically no difference to work on the astral and causal planes.

There could be some interferences, though, in the lowest energy 3d physical-etheric dark sector, when travelling in the region coincident with the troposphere (where practically all weather occurs,) as reported by the medium Hiralal Kaji. This is plausible as the number of dimensions of this sector is the same as the ordinary matter sector and energy levels are comparable. We also note that specific particles in this dark sector can convert to ordinary matter particles and back again. There is also a conjectured fifth force that may allow for weak interactions between ordinary and dark matter.

Liquid-Crystal Complex Plasma

As noted previously, complex plasma has the characteristics of a liquid-crystal. Our ordinary matter universe, and higher energy universes, are composed largely of complex plasma. This means each of these universes would have a liquid-crystal nature. Based on specific conditions, they may be predominantly liquid-like (or fluidic,) or crystalline in nature. Similarly, each dark plasmasphere would be liquid-crystal, and its structure could be fluidic or crystalline.

Nature of Light in Dark Plasmaspheres

Raymond Moody reports in his best-selling book "Life after Life" that near-death-experiencers (NDEers) meet beings of light who are not composed of ordinary light. They say, "they glow with a beautiful and intense luminescence that seems to permeate everything." Despite the light being much brighter than on the ordinary matter Earth, it doesn't hurt their eyes. The existence of these types of subtle light has also been recorded in religious and metaphysical literature over the centuries, including the Christian Bible, and Church records. Pope Benedict XIV admits, "There are hundreds of such examples to be found in our hagiographical records." Physicist John Cramer says that our universe could, without our knowledge, be superimposed on another "shadow" universe which has *its own light* and matter which does not interact with ours. Physicist John Schwarz (one of the co-founders of String theory) says that shadow matter would be essentially invisible to us because it wouldn't interact with the kind of light that we are able to detect.

Dark Light

Dark light or super light is the counterpart of the light that we are familiar with — composed of super photons. Each plane, brane, sphere, dark sector, or bubble universe would have its own kind of light. The speed of each type of dark light particle (in the vacuum) may be different (see chapter 11.) This is not surprising, since if each dark sector is viewed as a bubble universe in eternal inflation theory (also discussed in chapter 11,) each sector or bubble would have its own unique physical constants. This will pose additional complications in trying to detect dark matter particles from the relevant dark sectors, which probably has not been considered fully by researchers.

CHAPTER 9

❋

Dark Plasma Bodies

Biology and Appearance

The generic properties and appearance of dark plasma bodies, discussed below, arise naturally from the nature and behavior of complex magnetic plasma. A review of the literature relating to counterpart bodies of human beings, ghosts, angels, and religious figures in apparitions, will show that they display these generic features. The discussion below is a quick summary about the nature of these bodies. For more details, please refer to the author's book "Our Invisible Bodies."

Biological Processes of Dark Plasma Life-Forms

Dark plasma bodies are composed of complex plasma. Complex plasma has the characteristics of a *liquid-crystal*, which makes it superior to water in its ability to support life. These bodies would therefore be expected to display signature features of complex plasma, some of which are discussed below.

Filamentary Birkeland Currents

A dark plasma body generates a magnetic field to protect it from unwanted radiation in the ionized environment of the plasmasphere (just as the Earth's magnetic field protects the planet from harmful solar radiation.) Charged dark matter particles spiral around these magnetic field lines, as filamentary Birkeland currents. In metaphysics, these have been called "meridians" by the Chinese and "nadis" by the Indians. The dense network of currents along the magnetic field lines serve as the *skeleton* of the plasma body, giving structure to it. These field lines become illuminated when currents of charged dark matter particles course around them, generating dark electromagnetic radiation. These illuminated field lines has been reported by experimental metaphysicist and clairvoyant Barbara Brennan, a former physicist at NASA.

Vortexes

When filamentary Birkeland currents cross, they pinch or merge to form nodes. As the fast-moving dark matter particles in the filaments flow in different directions, they cause a circular swirling motion and generate angular momentum. This causes vortexes (or vortices), and cone structures, to form in the plasma, which becomes larger as they reach the surface of the dense humanoid figure within the ovoid. The motion of charged particles gives rise to intense magnetic fields (based on basic electromagnetics,) perpendicular to the plane of the vortex, which gets twisted due to the circular motion. The vortexes and cones are called collectively "chakras" (meaning "wheels") in the Indian yogic literature.

The rotating magnetic and electric fields generated by the circular motion are connected to a much larger electromagnetic field around the dark plasma body. This larger field attracts, and guides charged energetic particles from the environment along its magnetic field lines and into the vortexes in the body of the plasma life-form using helical paths. The angular momentum generated by the charged particles, plunging into the vortexes at different angles, and the resulting magnetic torsion, provide additional torque to the vortexes, making them rotate more rapidly.

The network of vortexes and filamentary currents function as the *circulatory system* in the dark plasma body. The vortexes suck in electromagnetic energy (in the form of charged energetic particles) from the environment and uses the network of filamentary currents to distribute electromagnetic energy throughout the body to power it. After the energy is transferred by the particles directly to the organs, and the excess stored in internal capacitors around the torso, they are discharged. These capacitors, or energy reservoirs, are well-studied in Chinese acupuncture and are called the "eight extraordinary meridians or vessels." A double layer, discussed further below, is a natural structure in plasma consisting of two parallel layers of opposite electrical charge. This structure is identical to a capacitor and can function as one.

The discharged particles are transported by the filamentary currents from various parts of the body, to be excreted through large vortexes (which serve as excretory organs,) as well as micro-vortexes (analogous to the pores of the skin.) In this way, the filament-vortex system also serves as the *excretory system*. The discharge through the micro-vortexes or pores, generates radiation, which is often described as a colorful aura around the body. There are many different types of vortexes in the dark plasma body, which come in different sizes and configurations.

Double-Layers

The humanoid body is enclosed in a plasma bubble, in the shape of an ovoid (discussed below.) This ovoid is enclosed in a plasma double-layer, commonly called the "auric sheath" in metaphysics. Together with the magnetic field and the bubble, the plasma sheath, which acts as *skin*, protects the dark plasma body from unwanted radiation and intrusions. Double-layers also enclose each concentric shell in the ovoid and also serve as capacitors, to store energy, within the dense humanoid form.

Appearance of Dark Plasma Bodies

Plasma Bodies frequently appear as Balls of Light

The natural stable shape of plasma bodies, within Earth's gravitational field, and without any artefacts (e.g., a temporary humanoid body,) has a tear-drop shape within Earth's gravitational field. In zero-gravity space, it will morph into a spherical ball. This is the most stable shape as it has the lowest possible ratio of surface area to volume and therefore requires the least amount of energy to maintain its shape. The shape changes are very similar to what happens to a candle flame (a near-plasma) on Earth, where it takes on a tear-drop or elongated ovoid shape due to gravity, and in zero-gravity space, where it takes on a spherical shape. We should not be surprised, therefore, if visitors from dark plasmaspheres (whether as religious figures, angels, or ghosts) appear as ovoid-shaped balls or orbs, either at the start or throughout the encounter. Other less stable forms may also appear, for e.g., vortices and vapor or a cloud of dispersed particles. Usually this is because they are less energetic entities or that the dark sight of the observer has not been developed, resulting in partial sightings of the body. Based on plasma dynamics, there will be concentric shells within the ovoid which are rotating (just like in a plasmasphere.) (This is indicative of a Birkeland current.)

Plasma Bodies Emit Light

Unlike the ordinary matter body, which is visible, mainly because of reflected light, plasma emits light because of "spontaneous emission." This is when charged particles fall to lower energies and emit photons. Furthermore, there are collisions between particles in the plasma bodies (including the ovoid as a whole) and high energy particles in the environment — similar to what happens in auroras and inside fluorescent lamps, as well as the arc discharges from a welding torch – which generates light and heat. This is why angels glow and have coronas! It is also the reason why in the astral world there is a diffused luminosity, not obviously coming from any special direction, as reported by Leadbeater. There are three modes of light emission in plasma – the dark mode, the glow mode, and the arc mode. The dark mode is when very little light is emitted, the glow mode is when a gentle light is emitted, and the arc mode is when a bright, intense light is emitted.

Electronic and Photonic Plasma Bodies

The lower energy plasma bodies are more matter-like and contain a lower proportion of photons. They are fiery in appearance. The higher energy plasma bodies are more radiation-like and contain a higher proportion of photons (i.e., particles of light.) They are light-like in appearance, and therefore sometimes described as "light-bodies" in the metaphysical literature. When traveling through a plasma, a photon gains mass due to the oscillations of the particles in the plasma, which determine the plasma frequency. This frequency is directly related to particle density and the strength of the charge, as well as the temperature, and inversely related to the mass of the charged particle. The higher the density of the plasma, the higher will be the mass of the photon. This will make the photons more matter-like. In a sense, it is analogous to the effect of the Higgs field which gives mass to elementary particles. In the ascending energy pathway, however, the density of dark plasma bodies and plasmaspheres fall significantly. Hence, the mass of the photon will continuously decrease, and the plasma body will become increasingly radiation-like.

Plasma Bodies Can Generate Ordinary Plasma

While dark matter particles interact only weakly with ordinary matter particles, there are certain processes that could occur during these phenomena that would allow observers in the ordinary matter universe to observe them indirectly. When dark matter particle densities are high, there is a greater probability of them colliding into ordinary matter particles. This will displace ordinary electrons, ionizing the ordinary matter. In other words, ordinary plasma will be generated, as well as ordinary light, heat, and electricity. This has been described as the "dark ionization process" by this author. Additionally, certain dark matter particles, including axions and dark photons, can convert to ordinary photons, generating light and heat, which may be visible to human observers, as well as to our instruments, including radar and a variety of cameras.

Plasma Bodies Generate Colorful Auras

Liquid-crystal plasma bodies are "thermochromic." These bodies will therefore display definite colors at specific temperatures. When a crystal made up of long rod molecules is heated, the crystal loses positional order and becomes a fluid. However, it may retain its orientation and form a threadlike liquid-crystal state. If the repeat length of patterns is of the same order as the wavelength of visible light, it can function as a diffraction grating and generate visible colors. Because the spacing changes with temperature, the material changes color with temperature (i.e., it is "thermochromic.") A similar process takes place in dark plasma crystalline bodies that undergo a transition to the fluidic state. It is well-known in metaphysical literature that the fluidic astral body can be very colorful.

Besides the colors of the dark plasma body, there is also a colored radiation that emanates from the bodies. In this book, we will confine the meaning of "aura" to this radiation. The auras are generated in a process not unlike the one that gives rise to the aurora borealis (or the "Northern Lights.") We have already noted that there is a tendency for charged particles to follow Earth's magnetic lines of force in chapter 7. In the process they collide with gases in the neutral atmosphere creating colorful auroras. Similarly, charged dark matter particles are captured by the dark electromagnetic fields of our dark plasma bodies and spiral around the magnetic lines of force, and into the vortexes in the bodies. In the process, they collide with particles in the environment, and within the ovoid, to generate the primary auras. Conversely, when these particles are discharged through the micro-vortexes or pores (as discussed earlier,) they generate secondary auras.

Plasma Bodies Can Alter their Opacity at Will

The degree of opacity of the plasma body can be changed through an act of will by the plasma life-form. If the plasma frequency is higher than the ambient light, the plasma will become like a reflective mirror (reflecting the light) and shine like a metallic object. If it is less, the plasma will become like a refractive lens, and be transparent to the light (in other words, it will disappear.) The plasma frequency is directly related to its density. So, if the plasma life-form wants to manifest, it will just need to increase its density. If it wants to disappear, it just requires reducing it. The bodies of angels and deities have been observed as being brightly shining, as well as translucent, in the metaphysical literature, for e.g., by Leadbeater and the seers at the Marian apparitions. Light travels at approximately 300 thousand km/s in a vacuum, which has a refractive index of 1.0, but it slows down to 225 thousand km/s in water and glass with a similar refractive index. Plasma has a refractive index when it acts like a lens and therefore the speed of light slows down in plasma. The higher the density of the plasma, the higher the refractive index, and the slower the speed of light. Light also slows down when photons gain mass in a plasma.

Plasma Bodies have Features Associated with the Sun

Dark plasma bodies have many features similar to the visible Sun — which is after all a hot ball of ordinary plasma. Coronal auras and discharges, granulation and spicules are all features associated with the Sun and dark plasma bodies (including angels, deities, and ghosts.) The bright coronal aura of the Sun and the bright auras of angels and deities are similarly produced. This coronal aura is even more pronounced for higher-energy angels and deities — such as in the public Marian apparitions, where the plasma is in arc mode. Spicules are short-lived phenomena, corresponding to rising jets of gas that move upward and last only a few minutes on the Sun. Microscopic spicules can be seen in the coronas of dark plasma bodies.

Plasma Bodies Can Pass through One Another

Leadbeater reported around 1910 that astral beings in the dark astral plane "can and do constantly interpenetrate one another fully, without in the least injuring one another." They "can and do pass through one another constantly, and through fixed astral objects." When passing through another plasma body for a short time, two astral bodies are not appreciably affected. However, there could be a certain resistance, similar to surface tension, caused by the thin plasma sheath around the ovoid. Furthermore, if the interpenetration lasts for some time, when the plasma bodies are moving at very slow speeds or are stationary, Leadbeater explains, they do affect one another as far as their rates of vibrations (or frequencies) are concerned. This is not surprising, as plasma bodies are dark electromagnetic bodies, and this represents a synchronization or entrainment of frequencies between two bodies. There could be changes in frequencies and distributions of charges over the plasma bodies after prolonged immersion. This is why devotees or disciples enhance the quality of their plasma bodies by sitting in meditation, immersed in the guru's dark plasma ovoid or halo.

The ability of plasma bodies to pass through other plasma bodies is a characteristic of low-density collisionless plasma. Leadbeater explains that particles in the densest astral matter are further apart than even ordinary gaseous particles. The inter-particle distances in ordinary plasma are typically one hundred thousand times that of ordinary solids and liquids. Particles in astral plasma would be even further apart. Hence, Leadbeater concludes, "it is easier for two of the densest astral bodies to pass through each other than it would be for the lightest gas to diffuse itself in the air."

Compactification due to the Ordinary Matter Body

Leadbeater has pointed out that 99 per cent of astral matter is concentrated within the periphery of the physical body. This suggests that there is an attractive force (a fifth force, which has not yet been identified by science) between dark (astral) matter particles and ordinary matter particles. This results in a correlation in the densities between dark and ordinary matter. (This is also borne out in astrophysical observations.)

Low-energy, as well as high-energy particles, are attracted to the dark plasma body. However, slower moving low-energy particles would succumb more readily to the effect of any attractive force and will therefore have a higher tendency to be compactified within the perimeters of the ordinary matter body, forming a (relatively) dense dark matter body within the ovoid. The presence of the ordinary matter body therefore forces the physical-etheric and astral bodies to be more compactified, while it is associated with it. Psychokinetic factors, such as having a strong self-concept, and constant identification with the ordinary matter body, will increase the degree of compactification. At higher energies, due to higher temperatures, the attractive fifth force will become progressively weaker. Hence, the degree of compactification of higher energy bodies, due to the effect of denser lower energy bodies, will taper off in the ascending energy pathway. Hence, there will be more mixing between lower and higher energy particles in higher energy bodies.

A person practicing intense meditation and prayer generates much ordinary heat (as often recounted in the relevant literature,) and also dark heat, which may be able to expand the volume and lower the density of the dark plasma body, thus counteracting the effects of any compactification. It has been observed by metaphysicists that, during sleep or in a relaxed mental state, when the dark plasma bodies separate slightly from the ordinary matter body, they expand (and the density reduces.) When they re-align on waking or because of an increase in stress, they contract. (This is similar to Earth's plasmasphere which expands and contracts under different conditions.) So, stress can also further compactify the plasma body. A relaxed mental state, which reduces stress, can therefore lower the density of the dark plasma body.

Composition of Dark Plasma Bodies

Leadbeater says that a person is constantly building his dark plasma bodies, accumulating, or releasing, coarse (i.e., low-energy particles) and/or finer (i.e., high-energy) particles, by his passions, emotions, and desires. He says, "If he is...to build himself a coarse and gross astral vehicle, habituated to responding only to the lower vibrations of the plane, he will find himself after death bound to the (lower-energy shells of the) plane during the long and slow process of the body's disintegration. He advises, "If he gives himself a vehicle mainly composed of finer material, [i.e., high energy particles,] he will have very much less post-mortem trouble and discomfort, and his evolution will proceed much more rapidly and easily." If we consider each biological cell as a particle, we can say that even the ordinary matter body is composed of many different types of particles which determine its health and longevity. As an analogy, if the owner consumes a lot of fatty foods, unhealthy cholesterol may accumulate, which may lead to an early death.

The nature and composition of the dark plasma body, in the same way, determines it's health and longevity, as well as psychological traits and desires. The trajectory and destination of the next life, when it occurs in the same ecological niche, is also determined by it. The composition of the body is the result of past actions, mental and emotional states. In this book, therefore, the nature and composition of the individual's dark plasma body (including its brain) is taken to be what religions would call "karma." (This approach is similar to Jainism, but the understanding of karma varies in different religions.) The brain (whether centralized or distributed) of the relevant dark plasma body will contain the neural correlates which determine mental and behavioral tendencies. These neural correlates are called "samskaras" in Yogic literature.

In higher energy particulate photonic spheres (discussed in chapter 17,) the body is a single particle. In this case, this particle would constitute what religions would call karma. As time dilates and the speed of light falls (as discussed in chapter 11,) the cycles of cause and effect slow down as we ascend to higher energy universes. Hence, the production and dissolution of the body (interpreted in a religious context as karma) slows down in the ascending energy pathway.

In addition to dark protons and electrons, the proportion of photons (light particles) in the composition of the dark plasma body will continuously increase as it ascends to higher energy shells and plasmaspheres. In the process, it will become increasingly radiation-like.

CHAPTER 10

❈

Dark Plasma Bodies

Communications and Transport

Communications

Plasma Bodies Emit Dark Electromagnetic and Acoustic Waves

Dark plasma bodies generate dark electromagnetic waves, as well as acoustic waves, which are technically known as "Alfvén waves," caused by the vibrations of magnetic field lines within the body. The dark electromagnetic radio waves can propagate more or less freely through a plasma from the source to a receiver and they become part of the remotely sensed electromagnetic spectrum of the body. These waves radiating out of dark plasma bodies contain much real-time information about the person — both his physical and psychological state.

Plasma Bodies Generate Thought-Holograms

Thoughts can modulate these electromagnetic and acoustic waves radiating out from dark plasma bodies to generate thought-holograms in the plasma environment, using processes similar to those used in current technologies such as cymatics and laser-induced plasma holography. Cymatics is a process in which acoustic waves vibrate to form three-dimensional images in a medium, such as water or sand. Laser-induced plasma holography is a process in which the air is ionized differently in different places by lasers, generating plasma with different properties, to create three-dimensional objects with different shapes, textures, and colors. They can be seen, as well as be heard (if they vibrate) and felt. (This is based on observations of actual current plasma holography applications today.)

As the thought-modulated dark electromagnetic and acoustic waves generated by a dark plasma body passes through the dark plasmasphere, they reverberate to create density and temperature fluctuations in the already ionized plasma environment. These fluctuations generate different colours, textures and shapes due to the nature of thermochromic liquid-crystal plasma (discussed in the previous chapter.) They form extended objects, which are generally known as "thought-forms" in the metaphysical literature but will be called "thought-holograms" in this book. If large groups of dark plasma bodies generate similar thought-holograms, these will clump, connect, and densify to produce holograms of larger objects and even landscapes – analogous to those produced in "holodecks" in Star Trek movies.

Plasma Bodies Can Communicate Telepathically

There are various methods of communication between dark plasma bodies. The list below is not meant to be exhaustive. The first method, for medium-range communications, is to use *acoustic* (or pressure) waves in plasma. This is similar to generating sound, except that it is a direct transmission of thoughts – as noted by Leadbeater. Plasma bodies, encased in ovoids, are immersed in low density plasma within the dark plasmaspheres. Vibrations caused by thoughts generate acoustic waves within the ovoid. When the ovoid vibrates, acoustic waves reverberate in the low density plasma environment. These vibrations are received by other plasma bodies in the vicinity, which vibrate in unison and decode the messages. This is confirmed by metaphysical observations. For e.g., Leadbeater explains, "Every thought of definite character [i.e., a concrete thought] radiates an undulation [i.e., a wave.] This rate of oscillation communicates itself to the surrounding matter in the same way as the vibration of a bell communicates itself to the surrounding air. This radiation travels out in all directions, and whenever it impinges upon another mental [i.e., dark plasma] body in a passive or receptive condition it communicates something of its own vibration."

Secondly, for long range communications, dark plasma bodies (being electromagnetic bodies) can emit electromagnetic waves (analogous to long range radio waves) which are modulated by the thoughts of the owner. Plasma is a good conductor of electromagnetism and electromagnetic waves, so the plasma environment in the dark plasmaspheres is ideal for transmitting and receiving these waves. Scientific researchers, Sanduloviciu and Lozneanu, have found that plasma spheres, generated in the laboratory, can communicate "by emitting electromagnetic energy, making the atoms within other spheres vibrate at a particular frequency." Plasma antennae can be used to transmit and receive dark electromagnetic waves. These devices are actually used in scientific laboratories today and are becoming increasingly popular due to greater functionalities. A living plasma life-form would be able to operate a plasma antenna that evolved as an organ within its body over millions of years. This is not surprising. Afterall, there are even more complex organs in the ordinary matter body, like the kidneys, lungs, and brains. From our current experience, the two methods used above may be interpreted as electromagnetic and acoustic wave telepathy - since plasma life-forms receive and send messages and thoughts without talking. The generation and transmission of these waves are very well studied phenomena in plasma physics.

Thirdly, for short-range or near-field communications, yet another method is for the dark plasma life-form to generate a thought-hologram (using mechanisms already discussed above) and direct it to the receiver. Thought-holograms would look like specific objects, for e.g., a figurine, a symbol or something similar. As reported by near-death-experiencers (NDEers,) messages can also be transmitted through the physical delivery of a "thought-ball." A thought-ball is like a zip file (to use computer terminology,) in the form of a floating plasma "bubble-drive." It is ejected and directed from one plasma body to another. The receiving plasma body absorbs, unpacks, and decodes the information.

Effect of Thoughts from Dark Plasma Bodies

Effect of Thoughts on One's Own Plasma Body

The brightness, intensity and colors of plasma bodies vary according to the properties of the plasma in their bodies — which in turn is the result of habitual emotional and mental states of the being — much like the way a PET scan or fMRI imaging of the brain shows colors in accordance with the neural activities in the brain — which correlate with psychological processes. These states are reflected in real-time in our dark thermochromic plasma bodies by changes in shape, brightness, contrast, colors, and opacity in different parts of the body, its aura, and the body as a whole. A metaphysicist or clairvoyant who observes and interprets the colors and other attributes of a dark plasma body is no different from a doctor diagnosing a scan or an image of the brain of a patient. Many "seers" attribute these colors to various mental or emotional thought patterns and behaviors — in the same way that neurosurgeons today view a brain scan in real-time and attribute emotional and mental states to the display colors, intensity, formations, and locations of these changes.

According to Leadbeater, surprise is shown by a sharp contraction of the plasma body, accompanied by an increased glow if it is a pleasant surprise and by a change of color to usually brown and grey in the lower part of the ovoid when it is not. Awe swells the devotional part of the body and the striations in the plasma become more strongly marked. Joy generates an additional brightness and radiance and produces ripples on the surface of the plasma body. General cheerfulness, in addition, produces bubbles and a calm serenity which is pleasant to see.

Effect of Thoughts on the Plasma Bodies of Others

We know that magnetic plasma generates an enormous amount of electromagnetic radiation — including those analogous to dark radio waves which can travel long distances. When changes occur within the plasma body, therefore, it is literally broadcasted to others over a wide range. Other dark plasma bodies can *tune* in and decode these waves. Once tuned-in, the plasma body of the other being can change its appearance in response to the first being's mental or emotional states — or the collective thoughts of a group of beings. These appearances can look very real. For e.g., the bodies can display a variety of clothes, skin, and hair, and also change shape, color, and its features within a blink of an eye. In Marian and other religious apparitions even the fine details of period costumes and hair can be seen.

The dark plasma body is using a technology that is similar to the generation of thought-holograms in the environment, except that this time it is within the body of the receiver. Thoughts from seers or observers modulate the electromagnetic and acoustic waves sent out by their dark plasma bodies. These waves create temperature and density fluctuations in the thermochromic liquid-crystal plasma bodies of the receivers, for e.g., the relevant deities or angels, which transform to conform to the appearance expected by the seers. The effect is so real; it is indistinguishable from "real" clothes and skin. Even images rendered in a 4K or 8K pixeled plasma TV would pale in comparison, due to a variety of factors, including the higher frequency of dark light. In an etheric double, the appearance and feel of the body would be almost identical to the ordinary matter body. Nevertheless, these changes in appearance occur mostly on the surface of plasma bodies — they are superficial (the internal organs are different from our ordinary matter bodies.) This is similar to an ultra-high definition TV, where images appear on a thin screen, but immediately behind it are the electronics and mechanical parts, which have nothing to do with the internal organs of the characters on the screen.

In this way, a being can appear as a deity or even a fork-wielding devil to a person who is actually broadcasting or projecting his or her mental images to others. If a dying person has preconceived ideas about who he or she will be meeting, any being with a plasma body that appears before him may acquire the appearance of the preconceived person (assuming that the receiving plasma being intentionally tunes into the waves emanated by the plasma body of the dying person.) To this extent, the dying person will be experiencing a hallucination. Generally, however, only the appearance and general ideas about the preconceived person are cloned onto the third party plasma body. Often, a discussion with the third party may throw up some anomalies. The propensity for deception by other plasma beings in higher energy plasmaspheres appears to be much higher than in our current ordinary matter sphere.

Effect of Thoughts in Lower Density Dark Plasmaspheres

The Moon's gravitational force is one-sixth that of Earth. Jump on Earth and you'll be two feet off the ground and then come straight down; jump on the Moon with the same force and you'll be twelve feet off the ground, and you'll be free-floating for kilometers. So, it is much more hazardous to jump on the Moon. Due to different environments, the same force will give rise to considerably different results. In the same way, the impact of thought-modulated dark electromagnetic waves in a dense ordinary matter environment will be weak or insignificant. However, apply the same force in a low-density dark plasma environment, such as in the astral plasmasphere, and the impact may send you reeling into an empty void or confronting a self-made ferocious monster (as described in books such as the *Tibetan Book of the Dead*.)

The thoughts of an observer can easily mould the surrounding matter in the dark plasmasphere. The stronger the emotion, the more forceful the waves. Consequently, it will be very difficult for you to detach yourself from what is being observed as you will be unconsciously shaping your macroscopic environment, quite quickly and in dramatic ways. The impact will astonish the first-time observer as all the matter around would seem to be imbued with a life of their own and animated. In the case of an accident in the ordinary matter sphere, Leadbeater says, the rush of emotion caused by the pain, or the fright would flame out like a great light in the relevant dark plasmasphere and attract the attention of dark plasma beings nearby. In accidents, therefore, victims may sense invisible helpers (or angels.)

This is due to the interactions of thoughts, via modulated electromagnetic and acoustic waves, with the matter and invisible dark electric and magnetic fields all around. The waves would cause density and temperature fluctuations in the dark plasma environment, generating transient plasma formations that can be seen by dark plasma beings. These formations are analogous to what plasma physicist, Anthony Peratt, had reported for ordinary plasma (see chapter 19.) This animated matter, which does not belong to any being specifically, but is a manifestation of our own collective unsettled thoughts, has sometimes been characterised as proto-life-forms. For e.g., Leadbeater describes it as "elemental essence."

In a sense, they are more like cellular automata and bots. It is therefore important to start controlling your thoughts in this ordinary matter Earth. Although it may have insignificant effects here and now, it will alleviate post-mortem distress (PMD) and prepare you better for your next destination. Meditative techniques will help you to do this.

Effect of Thoughts in the Imaginal Realm

Trillions of thought-holograms from individuals in religious communities have been created over many centuries. Thought-holograms emanating from dark plasma bodies, linked to highly dense ordinary matter bodies, are composed of heavier and denser dark plasma (as previously discussed, there is a density correlation between ordinary and dark matter.) Where these heavier thought-holograms resonate with other similar thought-holograms, they connect, clump, grow larger, densify, and sink to the lower regions of the plasmasphere. As they accumulate, they produce the dreamscapes of different types of classical hells in the imaginal realm (discussed in chapter 14,) which are zoned according to religions. Conversely, the lighter thought-holograms generated by individuals, levitate to the classical heavens in the imaginal realm (discussed in chapter 15,) including the lowest heaven in the physical-etheric plasmasphere which stages UFO/UAP phenomena (as discussed in chapter 18) and Marian apparitions (discussed in chapter 19.)

Transport

Plasma dynamics varies with the type of analysis and the nature of the plasma. The type of analysis requires us to consider whether we are interested in the behavior of an individual particle or a collection of particles. The nature of the plasma requires us to consider whether it is a neutral or non-neutral plasma. (We will assume we are always talking about magnetised plasma.)

When looking at a dark plasma body as a single particle, we will have to know whether it is neutral or non-neutral (i.e., charged.) If it is neutral, its dynamics will be largely governed by gravity, as a first approximation. If it is non-neutral (i.e., charged,) it will be largely governed by electro-dynamics. When looking at the behavior of large collection of particles in magnetised plasma, we will need to know whether its fluidic or crystalline. If it is fluidic, magnetohydrodynamics (or "MHD" for short) is usually used. It combines fluid dynamics with magnetism. It does not deal with individual particles but describes plasma as a continuous fluidic medium. These dynamics can produce filamentary structures representing the flows of charged particles, as well as vortexes, as discussed in chapter 7. If it is crystalline, a kinetic approach is usually taken, using statistical methods. This can be used when studying plasma crystals and the production of ordered concentric shells.

Dynamics of Dark Plasma Bodies in a Dark Plasmasphere

The overall *neutral* dark plasma body, which is released from the ordinary matter body, is subject primarily to gravity. It can be treated as a single neutral particle. The shells to which this body levitates or gravitates into are based on its equilibrium position in the plasmasphere. As a first approximation, this is determined primarily by the specific gravity of the plasma body, relative to the plasma in the shell. (The specific gravity is the ratio of the density of the plasma body and the density of the plasma in the shell.) When the specific gravity is "1" it is in equilibrium. A more general indicator of the specific gravity in a gravitational field is weight. If it is lighter in weight, it will rise to higher energy shells. If it is heavy, it will sink to lower energy shells. If it is the correct weight, it will stay in the shell. This is illustrated well in the metaphor found in the *Egyptian Book of the Dead,* where the heart (i.e., the soul) is put on one plate of a weighing scale, and a feather on the other. When the heart has the same weight as the feather, i.e., when it is in equilibrium, it will be able to stay at that level of heaven. However, if it is dense and heavy, it will sink. The lifestyle of the individual and the strength of his/her self-concept, among other factors, determine the density and weight of the dark plasma body (as discussed in the previous chapter.) These natural dynamics are supplemented by directions from guides and the social network in higher energy human-linked biospheres. They will provide support to the wandering dark plasma life-form in its journey through the shells in the plasmasphere if they think the being needs help.

Although the word "soul" is not clearly defined and is generally avoided in this book because of its ambiguity, many who take the traditional view may identify the migrating dark plasma bodies as "souls." It is often said and noted by metaphysicists and psychics, and in NDE research, that similar souls usually come together to form groups or families. This is actually due to plasma dynamics. Overall neutral dark plasma bodies with similar properties will resonate with similar frequencies and come together to form groups in cells, within the shells. This process would be assisted by a network of guides from higher energy universes, if required. The twentieth century mystic, Paramahansa Yogananda, calls these spaces "vibratory quarters," (as discussed in the previous chapter.) The properties of the dark plasma body contribute, as a whole, to the composite frequency or what metaphysicists call the "vibratory signature."

> People don't go to heaven because of their good deeds. After death, people gravitate into groups according to the rate of their soul's vibration. Our soul naturally fits in the level of heaven we have developed within us.
>
> Kevin Williams, NDE researcher and writer

Plasma Bodies Can Generate Strong Electric Winds

A strong electric wind can be generated by plasma bodies. This is a well-known process, caused by the collisions between charged particles (electrons or ions) in the plasma and neutral particles, which causes a blast of neutral gas. Electric winds, one fourth the speed of a typhoon or four meters per second, have been produced in a laboratory. This technology is already being used to make consumer products, such as electric plasma fans, which do not require fan blades. The collisions can also cause the temperature to rise and the air volume to expand suddenly. This generates acoustic pressure waves that can cause winds, as well as the sound of thunder (which has been heard in Marian apparitions and UFO sightings.)

Electromagnetic Highways within Shells

Within shells, the motion of dark plasma bodies (whose energy levels are within the range of the shell) may be directed by Birkeland currents which serve as "electromagnetic highways" in the plasmasphere. By changing the strength and distribution of charges in the body through an act of will, certain parts may be made non-neutral. These parts act like charged particles, accelerating the body forward within the current flow at very high speeds. In other words, when an astral body "flies off" to another location, it is driven along these currents subconsciously, using electric propulsion. In this way, it would appear to the traveler that he/she is moving by thought alone. It is analogous to the passage of a charged particle being accelerated around a magnetic field line. There are other methods for locomotion, for e.g., electric wind propulsion (as discussed above,) plasma propulsion using a blast of plasma from the vortexes, and photonic propulsion for higher energy photonic-plasma life-forms. The use of photonic vehicles for inter-planetary travel had also been mentioned by the mystic Yogananda in 1946.

It should not be surprising that such movements can be effected through an act of will and psychokinetic forces. In fact, every time you stand, sit, or walk, your ordinary matter body is responding to your thoughts. However, for new entrants to the plasmasphere, these methods of locomotion may need to be learnt, just like how a baby would need to learn how to walk or swim. Otherwise, you will be subject to and be swept by the currents, as experimental metaphysicist Leadbeater observed:

> On the astral plane there are many currents which tend to carry about persons who are lacking in will, and even those who have will but do not know how to use it.
>
> Charles Leadbeater, Experimental Metaphysicist, 1910

Until they learn how to use these currents, newly released dark plasma bodies would generally therefore be subject to other forces, mainly gravity. Other plasma life-forms, either natives or former humans who have stayed in that plasmasphere for a long time, will know how to use these currents. They may help novices, if required, to bring them to different locations using the currents. It will be difficult generally for plasma life-forms from one shell to go down to lower shells because there will be resistance from the plasmaspheric pressure and the surface tension from the plasma sheaths surrounding the shells. Conversely, new entrants to the shell generally would not have the required energy to go to higher shells using these currents. In a sense, the equilibrium position of the dark plasma body betrays its energy level. So, at the initial stage of entry, dark plasma bodies would primarily be subjected to gravity and directions by their guides.

Phasing into Different Shells

During the life with an ordinary matter body, the higher energy plasma bodies are locked into or linked to a specific shell in the ordinary matter Earth sphere (in this case the surface of the Earth.) Hence, a heavier dark plasma body, which would ordinarily sink below the surface of the Earth, may be artificially supported (by the ordinary matter body) to dwell on the surface of the Earth. Conversely, a lighter dark plasma body that would ordinarily levitate to higher shells may be artificially "anchored" to the surface of the Earth. Nevertheless, these plasma bodies, especially the higher energy ones, do have some freedom of movement to other shells of the plasmasphere, depending on their level of identification with the ordinary matter body, which determines the strength of the link or coupling. When the link to the ordinary matter body is severed during death or temporarily relaxed during sleep or meditation, and this identification weakens or ceases, there is much more freedom.

Leadbeater notes that when the high frequency astral (plasma) body has exhausted its attractions to one level (i.e., a shell within a plasmasphere,) the greater part of its grosser matter (from the outer shell of the ovoid) will fall away, and it will find itself in affinity with a higher state of existence (in a higher energy shell.) Its specific gravity, as it were, is constantly decreasing. It steadily rises from the dense to the lighter strata (i.e., shell,) "pausing only when it is exactly balanced for a time" — i.e., when it is in an "equilibrium" position in the plasmasphere. In other words, as particles are ejected from the plasma body, and the kinetic energy (i.e., temperature) of remaining particles rise, the volume of the body (and the ovoid) increases and the density falls. This allows the body to rise to a higher shell.

The ejection rate and particle density may vary not only due to physical factors but also to psychokinetic and social factors. For e.g., a person who has a strong "self-concept" (particularly in the identification of the personal self with his/her ordinary matter body, for e.g., in terms of his/her thoughts, feelings and activities) would not only attract heavier, denser, lower-energy dark matter, but also create forces that would compactify the particles in the humanoid plasma body — reducing its volume and increasing its density and weight (see chapter 9 for more details.) This brings the being down to a lower energy shell.

At a more fundamental level, different states of consciousness, including different emotional and mental states, will attract particles with different energies that will respond to different frequencies. Lower frequency states of consciousness will attract lower energy particles that will form a dense plasma. Higher frequency states of consciousness will attract higher energy particles which will form a less dense and more tenuous plasma. If the dark plasma body is composed of high-energy tenuous plasma, it will levitate towards higher-energy and lower-density shells, away from Earth's core. These are places where heavens are normally thought to be located in many religions. If it is composed of low-energy dense dark plasma it will sink to lower-energy and higher-density shells. The most extreme of these places are where hells are thought to be located in many religions.

Phasing into Different Spheres

Transmigration from one dark plasmasphere to another, using the ascending pathway, takes place when a lower-energy and higher-density plasma body dies, and the locus of consciousness shifts to a higher-energy and lower-density plasma body. This shift is mediated by an identity particle-wave, discussed in chapter 17. It is inevitable that an existence in a low-energy sphere will attract lower energy and denser plasma of that sphere. Hence, inhabitants of every sphere must remove the particles belonging to that sphere, before they move up to a higher energy sphere, through a detoxification process. This is the function of the so-called "hells." During life in the compactified body, meditation and prayer can generate heat which increases the volume and lowers the density of the body and the ovoid, by increasing the kinetic energy of particles, which also facilitates in expelling them.

Subjective Perceptual Changes

As the cognitive and sensory systems of one body deteriorate, the locus of consciousness shifts to the corresponding systems of the next higher energy body. This usually happens gradually and intermittently (as evidenced in pre-death experiences.) As the signals from the ordinary matter body fades out during its death process, signals from the higher energy body become stronger, with sense impressions from the next universe becoming more noticeable and vivid. So, gradually, almost imperceptibly, the person begins living and operating in a different world.

Just before death, he/she will start seeing the counterpart dark plasma bodies of his "living" family and friends intermittently, rather than their ordinary matter bodies, as the locus of consciousness oscillates between the two bodies. The same happens when phasing into each succeeding higher energy dark universe. Experts who have activated and developed their spectrum of dark plasma bodies, have greater liberty to switch back and forth between dark sectors. The phasing into higher-energy dark plasmaspheres will also bring with it a gradual awareness of additional spatial dimensions, and a dilution in the time dimension. This is discussed in the next chapter.

CHAPTER 11

*

M-Theory and Eternal Inflation

> No physical theory we know of dictates that there should be only three dimensions of space. Although we can't see them with our eyes or feel them with our fingertips, additional dimensions of space are a logical possibility. Do I believe in extra dimensions? I confess I do.
>
> Lisa Randall, Leading M-Theorist,
> Harvard University, 2005

M-theory – A Magical Mystery Tour

Physicists have realized that different superstring theories are in fact limiting cases of a single, more powerful theory — known as "M-theory." According to this theory, we live in a higher dimensional universe but are trapped within a lower dimensional "membrane" (or "brane," for short.) As a result, we are blocked-off from the rest of the multiverse. We can only interact with higher-dimensional branes through the combined gravitational fields of all the universes in the multiverse.

Earth's dark 3d-double is in the physical-etheric mirror universe, often discussed by metaphysicists. Under M-theory, it could be described as an "etheric 3-brane." It may even be conceived as an extension of our 3d-brane, which is folded over and pressed together with our brane. Imagine a sheet of paper as a brane, and that a body was lying on the sheet. We then fold the sheet, so that the fold is facing the rest of the sheet. Then we slide a duplicate (or double) of the body along the top of the sheet until it reaches the end. It will now face the original body, but upside down, with its head facing the feet of the original body. We then rotate the body, so that both bodies face the top. If we bring the sheets together, you will find that the duplicate (or double's) body would be left-right reversed, as well as top-down reversed. This was what was observed by experimental metaphysicist, Robert Monroe, about the physical-etheric body during an OBE (out-of-body experience) – as discussed in more detail in chapter 6. The 3d space of the physical-etheric mirror Earth is reversed both along the x-axis (i.e., horizontally,) as well as the y-axis (i.e., vertically,) and in the z-axis (with a background-foreground reversal.)

Metaphysicist Leadbeater observed in 1904 that there are many characteristics of the astral world which agree with remarkable exactitude with a world of four (space) dimensions. He also said around 1910, "Our minds can grasp three (space) dimensions only, whereas there are four (space) dimensions on the astral plane, and five on the mental plane." He also gave many details about these branes (or "planes" in the metaphysical literature) in his various books. In the context of mainstream physics, under M-theory, we would say that the astral universe is an "astral 4-brane." The "mental universe" would be a "mental 5-brane."

According to M-theory, the higher dimensional branes shape and structure our universe — an observation often noted in the metaphysical literature. The theory also requires that the standard particles, of which we are familiar, are *confined* to our universe. All ordinary matter and energy (including photons or light particles) must "stick" to the surface of our 3-brane universe — like open strings that are anchored to a surface. The extra dimensions in the multiverse could not be experienced because ordinary matter and light could not go there. Our light, radio waves, magnetism, quarks, electrons, all operate only on our 3-brane. Gravity, though, can propagate as closed loops of string that does not stick to a surface, enabling it to explore the extra-dimensional space.

M-theory accords well with the metaphysical evidence. The physical-etheric, astral, and higher energy dark matter particles are known to be *confined* to particular universes. It has been acknowledged by metaphysicists that astral matter, just like dark matter, is subject to gravitation. Metaphysical evidence also points out that we cannot experience extra dimensions unless we use a body composed of the relevant super-physical (dark matter) particles and exercise the cognitive-sensory systems associated with that body.

Eternal Inflation Theory - Bubble Universes

According to the mystic Paramahansa Yogananda, "the entire physical creation hangs like a *little* solid basket under the *huge* luminous balloon of the astral sphere" (emphasis added.) If we consider the "balloon" as a bubble, this suggests that the universes of ordinary and physical-etheric matter, like tiny 3d bubble universes, sit inside a much larger 4d bubble which is the dark astral universe, both interpenetrating each other. The astral bubble sits inside an even larger bubble – the 5d "mental" universe, which sits inside a larger bubble – the 6d "spiritual" universe. There are bubbles within bubbles. The general view of metaphysicists, and specifically Yogananda, is consistent with mainstream science and the standard cosmology model, specifically the eternal inflation theory.

Eternal inflation theory is an extension of the big bang theory in standard cosmology. According to this theory, local bubble universes individually nucleate in the high energy vacuum and inflate as vacua, analogous to bubbles of air in water. This triggers the creation of matter and energy within the bubble (i.e., the big bang relating to the bubble universe.) It then grows and evolves against an ever-expanding and larger background multiverse. Each bubble can have a different number of dimensions. It can also have different fundamental constants, which includes the speed of light, or more fundamentally, the speed of causality. The main proponents of this multiverse model include well-known theoretical physicists: Alan Guth, Paul Steinhardt, Andrei Lindei and Alex Vilenkin.

Alek Vilenkin explains in his book, "Many Worlds in One," that, according to eternal inflation theory, "Bubbles filled with lower-energy vacua nucleate in the inflating high-energy background, and still lower-energy bubbles nucleate inside of them." This means that lower energy universes will form within each bubble universe, so there will be bubbles popping-up within bubbles:

> The highest-energy vacuum will inflate the fastest. Bubbles of lower-energy vacua will nucleate and expand in this inflating background...The interiors of the bubbles will inflate at a smaller rate, and *bubbles of still lower energy will pop out inside of them*...We live in one of these bubbles. (Emphases added.)
>
> Alex Vilenkin, Leading Eternal Inflation
> Theorist, Tufts University, 2005

Furthermore, the earlier and higher-energy bubbles will inflate the fastest and expand at a faster rate than lower-energy bubbles that are formed later. If that is case, then they will reach a state with only dispersed particles, and essentially a void, before the lower energy universes. Andrei Linde, also a leading contributor to eternal inflation theory points out that each successive bubble universe that is born would be a lower energy universe and could have a different number of spatial dimensions, similar to branes.

> It seems likely that the universe is an eternal, self-reproducing entity divided into many mini-universes, with low-energy physics and perhaps *even the dimensionality of space differing from one to the other.* (Emphasis added.)
>
> Andre Linde, Leading Eternal Inflation Theorist, Stanford University, 2005

Each of M-theory's brane could be considered a bubble universe with a specific number of large dimensions, with the rest of the dimensions compactified.

The eternal inflation model is used as a template in this book, although the details of this theory are still being worked out and tested. Generally, the latest observational data of the cosmic microwave background supports this theory. The more refined model may probably see inflation proceeding in a less chaotic manner. Furthermore, the question of whether the multiverse was eternal in the past may become less relevant as we build quantum gravity models that teach us that space and time are not fundamental to reality.

Evolution of Time in Higher-Energy Universes

Based on the metaphysical literature, the duration of the present moment, or the "now," increases as we move up to higher dimensional plasmaspheres. What is perceived as the "present" in a higher dimensional universe occupies a longer period in the ordinary matter universe. For e.g., a year in a lower energy universe (such as the ordinary matter universe) may be perceived in a minute in a higher dimensional universe, and this minute may encompass both the future and the past of the lower energy universe. The present moment, instead of being point-like in our universe, becomes interval-like (i.e., from a zero-dimensional point to a one-dimensional line.) As we move up to higher energy universes, it expands to encompass more and more events, both in the past and future of the lower energy universe. As this happens, time becomes more space-like. (It is space-like in the sense that we increasingly have the freedom to move in any direction we want – both to the past and the future.) Ultimately, the time dimension would be no different from a spatial dimension. There is almost "pure space" in higher energy universes. This gradual disappearance of time is consistent with modern physics. For e.g., the influential loop quantum gravity theory does not require time to explain the universe at a fundamental level.

In 1899, metaphysicist Leadbeater reported that on a very high plane the past, present, and future all exist simultaneously. He gives the analogy of a passenger in a train who would consider the passing landscapes as successive and would be unable to conceive their coexistence through direct perception. However, people outside the train would see it as one continuous landscape, encompassing all the partial views of the landscape that the passenger sees inside the train. He apparently anticipated Einstein's "block universe" concept and the relativity of simultaneity (or the relativity of the present moment, the "now") through his direct experience with super-physical universes. The block universe contains the past, present, and future – it will be like looking at the whole landscape at once. Experientially, there is only space and no time dimension.

To examine more closely what happens when time becomes more space-like we can use a Minkowski diagram. In this diagram, space is seen as the x-axis, time, the y-axis, and the speed of causality (commonly shown as the speed of light) is seen as a line rising forty-five degrees from the origin, between the positive x and y axes. The metaphysical evidence, indicating that higher dimensional and higher energy universes see an increase in the duration of the present, means that time is becoming more space-like. In the diagram, the region that is "space-like" would then be seen to increase, and the region that is "time-like" will decrease. The region that is time-like represents a region where causality operates. If this region contracts, it means that the operation of causality is diminishing. Geometrically, this is represented by the current forty-five degree line in the diagram, representing the speed of causality, rotating counter-clockwise, from the x-axis (space) to the y-axis (time).

If the speed of causality decreases as we ascend to higher-dimensional and higher-energy universes, it means that the speed of light would decrease. This may have effects on other constants in physics. The fine structure constant is a well-known constant in physics. It is directly correlated to the strength of the elementary charge, and inversely correlated to the speed of light. It has been observed and conjectured to increase at higher energies. If that is so, then the increase in the value of the fine structure "constant" could be attributed to a decrease in the speed of light in each succeeding higher energy universe. Conversely, this means that the speed of light increases in each succeeding lower energy universe. Furthermore, an increase in the fine structure constant in each succeeding higher energy universe means that it will be increasingly difficult to form atoms and impossible to generate carbon-based life – hence, the state of matter will be predominantly plasma and the most probable life-forms would be plasma and photonic-based.

The Local Multiverse Model

Superstring theory implies a ten-dimensional world, with nine space dimensions and one time dimension. M-theory implies an eleven-dimensional world, but one of the spatial "dimensions" is zero dimension or a point (meaning no extension.) Ignoring that, we would have a total of ten dimensions, including the time dimension. The author believes, however, that the universe/multiverse did not start with a time dimension. It started with ten spatial dimensions. Then, one spatial dimension evolved incrementally into the time dimension that we experience, in each succeeding lower energy universe, as the speed of light rose. We will use these considerations to construct a basic minimal model of evolution for the local multiverse.

The author suggests the following scenario, combining insights from eternal inflation theory and M-theory, and taking the descending energy pathway: After the first highest energy 10d bubble universe formed in the high energy vacuum, it reproduced smaller and lower-energy energy bubbles of vacua, which then grew universes within them. The energy density of the vacuum inside the smaller bubble is lower than the larger bubble within which it is situated. This creates a border or wall between the vacua of the two bubbles. The energy density of the vacuum in each new bubble universe would keep falling until it reached the very low energy density of our 3d ordinary matter universe. Based on computations of the theoretical energy density of the quantum vacuum in quantum field theory, this means that the energy density of the high energy quantum vacuum (from which the local multiverse emanated) could be about 120 orders of magnitude higher than the quantum vacuum in our ordinary matter universe. If we had eight main levels of universes at a lower energy (as discussed below) it would suggest (as a first approximation) that each new bubble universe fell in energy density by about 15 orders of magnitude.

There was a cascade of new bubble universe creations with its own corresponding big bang. Soon after the first bubble universe with ten spatial dimensions was born, one of the spatial dimensions started morphing into a time dimension. This generated a new bubble universe with nine extended spatial dimensions and one time dimension. Subsequently, analogous to a waterfall, each successive new bubble universe fell to a lower energy level, with a faster speed of causality and speed of light. Due to the lower energies, the new bubble universe lost one of its spatial dimensions, as it was not able to grow into a large dimension. This is analogous to the stump of a limb that does not grow after birth. In M-theory and superstring theory it would be said to be curled-up or compactified.

From the 6d to 3d bubble universes, each new universe splits into two: a universe with its mirror or shadow universe. They are the two E6 components of the E6xE6 gauge symmetry group, under superstring theory.

The symmetry between the two E6 components breaks at the birth of the universe, when gravity splits apart from the other forces of nature in the first second of the relevant big bang, so a slightly lower energy shadow or mirror universe is created within the other component universe. This mirror universe inverts or flips all the symmetries of the initial universe, relating to charge (C), parity (P), space (S), time (T) and matter-force (M). (The number of dimensions in the mirror universe will be the same as the initial universe.) Considering the liquid-crystal nature of complex plasma, each mirror universe would also invert its propensity to be predominantly fluidic or crystalline. Each pair of strongly coupled universes, would consist of one universe being fluidic, and the mirror being crystalline.

The "E6" twin of our ordinary matter Earth is the higher energy 3d physical-etheric double. (It is arbitrary which of the universes is called the "mirror" – it depends on the direction of travel. In the descending energy pathway, the mirror universe will be at a slightly lower energy level. In the ascending energy pathway, the mirror universe will be at a slightly higher energy level.) As each new lower energy bubble universe and its mirror is born, one of the six spatial dimensions of the E6 component will become compactified. All these universes (including the ordinary matter universe) make up about 30 per cent of the matter in the local multiverse/universe.

Oppositely charged particles, including dark protons and electrons, are created from the 6d universe onwards, producing self-interacting dark matter. In total, all the self-interacting dark matter in the dark plasmaspheres forms 15 per cent of the matter in our measurable local multiverse/universe. In the descending energy pathway, the density of each dark plasmasphere increases. As a first rough approximation the mass of the 6d, 5d, 4d and 3d dark plasmaspheres would be 1 per cent, 2 per cent, 4 per cent and 8 per cent of the total matter in the local multiverse/universe, respectively. The bulk of dark matter, comprising of neutral, slow-moving, and classically non-self-interacting particles, exist within four bubble universes - from the 10d to the 7d bubble universes. This makes up 70 per cent of all matter in the universe/multiverse. Dark matter was created in the higher-dimensional bubble universes, and the 3d dark bubble universe, before the big bang relating to our ordinary matter universe. The matter in these dark universes cast their shadow gravity during our universe's big bang, resulting in the gravitational lensing of light that is now seen as the leftover radiation in the cosmic the microwave background (CMB).

Each mirror universe has the arrow of time pointing in the reverse direction, with its spatial dimensions also reversed. The inhabitants within the bubble universe would experience time moving forwards. They would have the same origin in time as the twin, and flow to the future in equal time with the twin, with entropy increasing. However, an external observer of the two universes will measure that the arrows of time of the universe and its twin are pointing in opposite directions, and the space is reversed.

If the locus of consciousness is transferred from the body in one universe to its counterpart double in the mirror universe, both bodies will be at the same radius from the origin of time in those universes. Since time flows at the same rate, the amount of time experienced by the body that elapses into the future of each universe would be the same. When the locus of consciousness returns to the first body, it will land into the future of the first universe, after undergoing a reversal or inversion in the dimensions of spacetime.

The bubble universe and its twin or mirror universe interpenetrate each other and share the same spacetime landscape. There is therefore a "mirror-like" process in the origin of the pair of the universes, not a naïve reflection, like in a 2d mirror. From the perspective of M-theory, we can visualize eight folded branes that would represent the initial universe alternating with its mirror in each fold. Although only 1 time dimension is recognized, its structure is different in each universe.

In total, there would be four pairs (the universe and its mirror) of eight bubble plasma universes, and four very low density bubble particulate photonic universes (discussed in chapter 17.) These universes will form our local multiverse or lineage, all interpenetrating and nested in each other. Each bubble universe would last for trillions of years as it peters out to nothing. We would have eight cosmic generations in our lineage (each universe and its twin mirror universe is considered one generation.) The higher-energy universes serve as the parent, grand-parent, and ancestor universes of our ordinary matter universe. When small regions of these larger bubble universes intersect in the ordinary matter universe, they warp its spacetime fabric to generate the convoluted cosmic web of dark matter, a mesh or netting of several webs of dark matter, corresponding to each interpenetrating higher energy bubble universe.

CHAPTER 12

❋

Bodies and their Universes

Ordinary Matter Bodies

Scientists estimate there may be up to ten million species on the ordinary matter Earth. Each species has a body that is different from the next. The reality that the body presents subjectively, through its cognitive-sensory systems, to the life-form can therefore vary significantly.

Earthworms and caterpillars have "eye spots," that see only light and dark. Dogs have vision similar to a human who is red-green color blind. Colors that would appear very rich to us are pastel-like to cats. To the dog and cat, what appears red to us is dark to them; and what appears to be green is white. Many animals, on the other hand, exceed the ability of human senses significantly. For e.g., they may see many more colors than we do. We have only three photoreceptors, but the mantis shrimp has twelve, and pigeons, six. Bees and butterflies see ultraviolet light, and snakes see infra-red light, which humans can't.

These differences only relate to vision. There are many more sense modalities – sound, taste, smell, touch, pain, pleasure, and the list goes on. Not only are sensory systems different, brains and cognitive processes also differ, thus adding to the multiplicity of realities that are generated. All these numerous examples (and there are millions more) relate to differences in the experiences of subjective reality of only life-forms that are composed of ordinary matter and standard particles. What about life-forms or bodies that are composed of dark matter and exotic particles? Furthermore, how would we perceive a reality that had more spatial dimensions?

Hooper and Teresi, after interviewing almost two hundred brain scientists, asked, "If our brains were a different size and shape, what would our religions be like? If we had three brain hemispheres instead of two, would our philosophies, our geometries, our mythologies, our notions of causality, space, time, and number be radically different?" Recent theories in science can shed some light.

Insights from Science

M-theory (as discussed in the previous chapter) makes two important observations about a multiverse: firstly - the elementary particles and fundamental forces of a higher-dimensional universe (or "brane") do not interact with the particles or forces in a lower-dimensional universe (the only exception being gravity,) secondly - extra spatial dimensions can be infinitely large.

Additionally, many scientists today, do not consider space and time as fundamental. It emerges from the set of particles relevant to specific universes. In fact, the central finding of loop quantum gravity (an influential mathematical theory to unify quantum mechanics with Einstein's general relativity) is that space is not infinitely divisible and therefore not continuous. It is granular in structure and, just like matter, is made up of elementary "space particles" that are a billion-billion times smaller than the smallest atomic nuclei. Gravitons will mediate between matter particles and space particles to manifest the gravitational "force," to tell spacetime how to curve, warp or ripple – just as photons mediate electromagnetic interactions between charged particles. These gravitons, according to M-theory, will be free to move to other branes. The matter in higher-energy branes or bubble universes will therefore cast a gravitational shadow on lower-dimensional universes, including ours.

The author proposes that these particles of space are stuck to the relevant branes and universes, just as its matter and light are. The distinction between space and matter therefore evaporates. Just as the inhabitants of the brane or universe can only see the matter of the brane, they are also subject to its space, with a specific number of dimensions. The idea of time that is independent of matter also evaporates. The passage of time is regarded as internal to a specific universe and emerges out of the relationships between quantum events within the space of that universe. In other words, time emerges out of space – as discussed in the previous chapter. It does not have an independent existence and increasingly morphs into a spatial dimension as energy levels rise. The number of spatial dimensions we perceive is constrained by the energy density of the universe we inhabit and the type of particles of space that make up our universe (or brane) and our bodies.

> We live on one of the bubbles, which is an expanding three-dimensional spherical brane. *For us, this brane is the only space there is. We cannot get out of it and are unaware of the extra dimensions.* (Emphasis added.)
>
> Alex Vilenkin, Theoretical Physicist, Tufts University

It logically then follows that you will need to have a cognitive-sensory system, composed of the same elementary particles as a specific universe, to see the matter and light of that universe, as well as any extra, large dimensions. Even if the state of physics evolves significantly to be able to manipulate non-standard particles, we will not be able to perceive these realities directly, using our ordinary matter bodies or scientific instruments. This will be a natural boundary to empirical science within the ordinary matter universe.

What you can see, or measure, is constrained by the type of particles your body and scientific instruments are composed of. While science has been very successful in the ordinary matter sphere, future scientists will have to take up advanced meditation, and use their higher energy bodies and related instruments, to empirically test their theories of universes and multiverses – like what contemplatives have been doing for the past few millenniums, but in a more institutionalized and systematic way.

Limitations of Our Cognitive-Sensory Apparatus from the Metaphysical Literature

In 1892, Sir William Crookes posted an article in the "Fortnightly Review," where he states that it is not improbable that other sentient beings have organs of sense which do not respond to some or any of the "rays" to which our eyes are sensitive but are able to appreciate other vibrations to which we are blind. He added that such beings would be living in a different world to our own. In 1893, experimental metaphysicist Annie Besant talked of interpenetrating universes. She explained that if the entities in these universes did not have organs of sense like our own, if their senses responded to vibrations different from those which affect ours, "They and we might walk side by side, pass each other, meet each other, pass through each other, and yet be never the wiser as to each other's existence — an unconscious co-existence of intelligent beings," — echoing typical descriptions of shadow universes given by scientists today.

Around 1910, experimental metaphysicist Leadbeater explained that "a man living in the physical world sees, hears, and feels by vibrations connected with the physical matter around him. However, he is surrounded by (super-physical) "astral" and "mental" worlds, which are interpenetrating his own denser world, which he is normally unconscious of, because his senses cannot respond to the oscillations of their matter." He elaborated that to examine a (physical) object a man is using a physical organ (the eye) which is capable of appreciating only certain "rates of undulation" radiated by certain types of matter. (In other words, it interacts with only matter composed of certain types of particles, relevant to a particular universe.) If he should develop "astral consciousness," he says, he would then be employing an organ which is capable of responding only to the vibrations radiated by another and finer part of that object. If in developing the astral consciousness he had lost the physical — that is, if he had left his physical body — he would be able to see only the astral and not the physical. Hence, we would require the relevant body, composed of the relevant matter relating to a specific universe, to see objects in that universe.

The above views relate to the sensing of matter and force particles from other universes. But what about spatial dimensions? Leadbeater says that for us, "All dimensions beyond the three are to us as though they did not exist...physically [i.e., in the ordinary matter universe] we see only three dimensions — hence, we see all objects and beings only partially. Our lack of perceptive power, however, does not in any way affect the objects themselves." In 1910 he lamented:

> We find ourselves in the midst of a vast universe built of matter of varying degrees of tenuity, which exists in a space of (let us suppose) seven dimensions. But we find ourselves in possession of a consciousness which is capable of appreciating only three of those dimensions, and only matter of certain degrees of tenuity.
>
> Charles Leadbeater, Experimental Metaphysicist, 1910

To summarize the metaphysical literature above, we will not be able to directly sense phenomena in an alternate universe with the cognitive-sensory systems of our ordinary matter body, or directly measure them with our ordinary matter scientific instruments. What they noted is consistent with the latest mainstream theories of the multiverse: M-theory, as well eternal inflation theory. According to Leadbeater the higher dimensional plasmaspheres are "both around us here and now, yet so long as our consciousness is focused *inside the physical brain*, we are blankly unconscious of them" (emphasis added.) It is important to note that he is referring to consciousness relating to the ordinary matter brain. When the ordinary matter and physical-etheric body dies, the locus of consciousness is transferred to the astral body (and brain,) and at once we find ourselves seeing the astral part of our world, after having lost sight of ordinary matter reality. Later, when we lose the astral bodies, we begin to live in an even higher energy body. Then we become conscious of an even higher energy plasmasphere; but would lose sight of the lower energy astral and ordinary matter spheres. Hence, the plasmasphere or universe you would have awareness of depends on having a body that is composed of the same elementary particles that makes up that plasmasphere or universe.

As the elementary particles that the body is composed of changes, the cognitive capacity increases as we move up to higher energy universes, correlating to a specific state of consciousness. For e.g., NDE research suggests that the physical-etheric body has a much higher cognitive capacity than the ordinary matter body. This capacity is the computational power of the relevant brain (whether centralized or distributed.) More specifically, it is the number of computations it can carry out per unit of time. The larger cognitive capacity of higher dimensional dark plasma bodies can be attributed to several factors, among others. Firstly, the brain would have a higher-dimensional neural architecture, hence, it would allow for more nodes and connections between the nodes in the network. Secondly, cognitive capacity is boosted in higher energy universes as they become increasingly quantum-like at macroscopic scales. The number of computations, per unit of time, of a brain which has access to quantum computing far outstrips any classical computing.

As we ascend into higher energy bubble universes, the frequency and bandwidth of the cognitive-sensory system of each dark plasma body rises, as energy levels rise. This includes the cognitive frequency, i.e., the frequency of thoughts and emotions transmitted and received; and the sensory frequency, i.e., the frequency of sensory impulses received. For e.g., in terms of cognitive frequency, if the lower frequency emotions are taken to be romantic love, joy and a feeling of ease; the higher would be universal love, bliss and a deep sense of peace. If the lower frequency cognition produced concrete thoughts (i.e., composed of actual images,) the higher would produce abstract thoughts. In terms of sensory frequency, lower frequency sensations, such as smell and taste, would be replaced by a wider bandwidth of higher frequency audio and visual sensations in a higher energy universe, including sounds and colors that cannot be imaged or imagined in the ordinary matter body. Hence, thoughts and sensations become increasingly subtle and positive. Each brain in each higher energy body provides the relevant neural correlates for the specific states of consciousness accessible to that body. The frequencies and capacities of each body differs so enormously that near-death experiences or NDEers find it impossible to recall solutions and insights in their ordinary matter body, which came so easily while in their higher energy body. Negative thoughts and emotions only exist in lower energy bodies and spheres, where perspectives are narrower due to lower cognitive and sensory frequencies and capacities.

Limitations of Our Cognitive-Sensory Apparatus in the Scientific Literature

What Leadbeater pointed out in 1910 is exactly what leading theoretical physicists are concluding now. Mathematical physicist, Bernard Carr, notes in his talk given at the Euro-PA Conference in November 2003, "Our physical sensory systems reveal only a very limited aspect of reality." Lisa Randall, a leading M-theorist, says, "Our senses register only three large dimensions, so an infinite extra dimension might sound incredible." Michael Duff, another leading superstring, and M-theorist, says "if our senses are to be trusted, we live in a world with three space and one time dimensions. However, the revival of the Kaluza-Klein idea, brought about by supergravity and superstrings, has warned us that this may be only an illusion." Modern physicists seem to be practically paraphrasing what metaphysicists wrote almost a century before:

> We don't see extra dimensions directly, so everything should appear to us as if it is four-dimensional [including time.] Just as Flatlanders, who see only two spatial dimensions, could observe only two-dimensional disks when a three-dimensional sphere passed through their world, we can only see particles that look like they are in three spatial dimensions, even if those particles originated in higher-dimensional space*.
>
> Lisa Randall, Leading M-Theorist, 2005

* The author refers to the observation of higher-dimensional particles in three spatial dimensions as the "3d-correlate" of the higher dimensional particle. Randall's analysis, however, is incomplete. There is an assumption that the disks (the 2d correlates of the sphere) are composed of ordinary matter particles and therefore the Flatlander can actually see the disk. However, if the sphere was made of elementary particles which do not interact with particles of Flatlander's universe, the Flatlander will not be able to see or measure anything. So, there are actually two matters to consider with respect to being able to directly detect phenomena from an alternate universe: firstly, the nature of the particles and forces, and secondly, the nature of the dimensional structure.

Leading M-theorists have found that two very different theories which are constructed using different numbers of spatial dimensions can be equivalent. Juan Maldacena first conjectured such a relation in 1997 for a 5 dimensional universe. It was later confirmed for many other universes with different numbers of dimensions by Edward Witten of the Institute for Advanced Study in Princeton, N J, and Steven Gubser, Igor Klebanov, and Alexander Polyakov of Princeton University. Examples of this correspondence, usually called the "AdS/CFT correspondence" or more colloquially the "holographic principle," are now known for universes with a variety of dimensions.

Jacob Bekenstein, a leading theoretical physicist who was awarded the Wolf prize for discovering the holographic nature of black holes, says that creatures living in one of these universes would be incapable of determining if they inhabited a 5d universe of strings (described by string theory) or a 4d universe of point particles (described by quantum field theory.) He believes that the *structures of their brains* might give them an overwhelming prejudice in favor of one description or another. Effectively, this means that human brains are biased to reduce higher-dimensional space into a 3d correlate i.e., a universe of three dimensions, with different physics.

According to the Greek fable, any object that Midas touched turned into gold. Similarly, when classical objects measure quantum objects, they turn into classical objects. The author likes to call this the "quantum Midas touch." It betrays the use of a metaphorical pair of spacetime goggles (or VR (virtual reality) headset) whenever we interact with quantum reality. When conscious agents in a quantum reality use these goggles, quantum objects transform into classical objects, within a spacetime grid with a specific number of dimensions.

There is therefore a 'relativity of dimensionality.' So even if the astral universe is known to have four spatial dimensions, someone who just enters it (and whose astral body is linked to the ordinary matter body and brain) may perceive only the 3d-correlate of that universe. These two different observations, however, can be correlated and reconciled mathematically – just as travelers moving at different speeds, and experiencing different rates of time, would be able to reconcile the spacetime interval between themselves, under Einstein's Special Theory of Relativity. The 3d-correlate would have all the information in the higher-dimensional universe, just as a two-dimensional plate containing holographic encoding has all the information about a three dimensional object, which allows it to project it as a 3d hologram.

There are therefore three levels of interactions with higher dimensional particles or objects. Firstly, in the ordinary matter universe, using our ordinary matter body or scientific instrument, we will not be able to directly interact with particles or objects from another higher-dimensional bubble universe (except through gravity if the object was an astronomical body.) Secondly, if we could activate the relevant linked dark plasma body, then, on initial entry into the corresponding higher-dimensional universe we will be able to see and partially interact with the higher-dimensional particle, object, or astronomical body, but we will see it as three dimensional (i.e., it's 3d-correlate) and with a different physics. Thirdly, after prolonged immersion in the higher-dimensional universe, which would help us to weaken the link between our higher energy bodies from our ordinary matter body, we will then be able to see or interact directly with the higher-dimensional particle, object, or astronomical body over all relevant dimensions of that universe, using a different model of physics.

The newly arrived being to a higher-dimensional plasmasphere would therefore need some time to learn to sense the extra dimensions. This is similar to the experience of a newborn baby, which only sees in two dimensions initially for several months before it starts seeing in three dimensions. This phenomena was well-known by metaphysicists. In 1896, experimental metaphysicist, Annie Besant, explained that people must *learn* how to see astral objects. After some time, and only after mistakes caused by cognitive biases are corrected, she said, can the vision of astral objects (in the 4d dark astral universe) become more accurate.

While some aspects can be learnt, at a more fundamental and basic level, the body will need to be composed of the same elementary particles that the relevant brane (or universe,) is composed of, in order to interact with objects in them. The cases where those who are born blind can see during near-death experiences illustrate this well. It suggests that they are using the cognitive-sensory apparatus of an alternate body composed of different elementary particles which can interact with the light from the relevant alternate universe. As an analogy, each camera (infrared, ultra-violet, 0r visible) in a telescope interacts very differently with the light from a galaxy and shows very different images.

Why are Experimental Metaphysicists Able to See Extra Dimensions?

How do experimental physicists, including Leadbeater, Besant, Yogananda, Brennan and numerous others, able to see what most people do not? How could they perceive macroscopic objects composed of higher energy particles and extra dimensions? As the metaphysical literature asserts, this is because they used the cognitive-sensory systems of their alternate linked invisible (i.e., dark) bodies composed of elementary particles that belonged to these alternate universes. These bodies, and their universes, are not directly visible to the cognitive-sensory system of the ordinary matter body – as the two sets of particles from different universes do not interact – as required by M-theory.

However, these invisible bodies are linked to the ordinary matter body. Consciousness, like gravity in M-theory, propagates through all universes (or branes.) The locus of consciousness, mediated by the activation of an "identity particle," can therefore disappear from one body, in one universe, and appear in another body, in another universe. The body is a temporary vehicle for non-local consciousness to interact in a local frame of reference. It is analogous to a VR (virtual reality) headset that allows the observer to experience different realities. This non-local consciousness extends beyond the brain and can access memories via the identity particle - although this ability may be limited for the majority of humans without any training, such as the practice of meditation. Each identity particle is a different excitation of a single closed string (called the "Sutratma thread" in Hindu metaphysics.) Closed strings are inter-dimensional as they can explore extra dimensions in M-theory. This string plays the role of an inter-dimensional "DNA," which is also popularly known as the "thread of life." It connects the highest energy bodies to the lowest-energy bodies.

While scientists can only indirectly detect invisible dark matter halos around galaxies through gravitational lensing, the Hindu mystic, Yogananda, can see them directly as colorful auras:

> The divine dispersion of rays poured from an Eternal Source, blazing into galaxies, transfigured with ineffable auras.
>
> Paramahansa Yogananda, 1946

The auras seen by metaphysicists are the dark matter halos of science – not only around galaxies but also around the human body.

CHAPTER 13

❊

Dark Earths

Earth's Dark Plasmaspheres

The dark plasmaspheres or simply "Dark Earths," were discussed in chapter 8. These are the "super-physical counterparts" of Earth in the metaphysical literature. Since Dark Earths have mass, would we not feel their gravitational effects in the Solar System? Scientists have noted a number of anomalies relating to the accelerations of spacecraft and satellites in the Solar System. Our Sun is moving faster around the Milky Way than what Newton's formula would dictate, suggesting that it is under the gravitational influence of dark matter in the galaxy. However, the mass of dark matter within the Solar System is estimated to be low, and the planets in the Solar System comply with Newton's law. If there were Dark Earths, should the gravity emanating from this additional mass be more evident?

Gravitational Fields of Dark Earths

In large scale structures, such as the Milky Way, there is a small volume of ordinary matter sitting inside a huge volume of dark matter. The mass of the dark matter is many multiples of the ordinary matter. In this case, ordinary matter would be subject to the gravitational mould imposed by dark matter. However, if we compare equal volumes of dark and ordinary matter, the mass of the dark matter would be much less as its particle density is much lower than ordinary matter. The mass of dark matter in the Solar System is estimated to be low. Hence, dark matter will be subject to the gravitational influence of ordinary matter. Furthermore, since the mass is low, the effect of the mass of any Dark Earths on our measurements of gravitational fields will be low.

Additionally, the gravitational force generated by higher dimensional astronomical bodies, in higher dimensional universes, would propagate, not based on the inverse-square law, but the inverse-cube, inverse-quartic, or inverse quintic law and higher. The gravitational force would be spread out over many more dimensions and therefore will be weaker and shorter-ranged. Furthermore, not all of this gravitational force will be felt in our three dimensional universe. We will only be able to measure the force from gravitons that interact with the large dimensions of our ordinary matter universe. The rest of the gravitons will seep into and be trapped within the microscopic compactified dimensions of our universe, which superstring theory postulates would exist. The gravity from higher dimensional Dark Earths would therefore be proportionately weaker, the higher the number of dimensions. Conversely, the gravity from the lowest dimensional Dark Earth would be felt more.

Most of the shadow gravity would therefore be contributed by the lowest energy 3d Dark Earth (i.e., the physical-etheric Earth in the mirror universe,) which has the same number of spatial dimensions as the ordinary matter Earth. Gravity, emanating from this 3d Dark Earth, would therefore follow the inverse-square law, as in the ordinary matter Earth. However, most of the mass of the dark matter in Dark Earths, which would be subject to the gravitational field of the ordinary matter Earth, would be confined to a region below the crust of the Earth. This is in accordance with what metaphysicist Leadbeater observed, as well as scientific theory (more specifically the Lambda CDM (Cold Dark Matter) model,) and theorized density correlations between ordinary and dark matter – discussed previously. Physicist David Peat estimates that about 10 per cent of Earth's core is dark matter. The bulk of the weak gravity from this dark matter, below the crust of the Earth (and Moon,) has already been considered in our current estimates of gravity, as it has been conflated with the gravity generated by ordinary matter.

Despite a low mass, and low density, halo of dark plasma, interpenetrating the Earth, Dark Earths could still support a plethora of low mass, highly energetic, radiation-like plasma-based life-forms. The invisible spaces between the widely dispersed particles in a low density dark plasma are not empty – they contain dynamic electric and magnetic fields which can create complex structures.

Dark Earths

Metaphysicists have long known about the existence of these Dark Earths, existing in different universes, which interpenetrate the ordinarily visible physical Earth; and have given them names, as follows (in ascending order of their energy levels and dimensionality): Earth's 3d-double or the "physical-etheric" sphere, the 4d astral and other higher dimensional spheres (called by various names, including "mental", "spiritual," and "causal.") Access to each Dark Earth gives us access to the huge (and possibly infinite) universe that they inhabit.

If there are Dark Earths, in the form of low density dark plasmaspheres, it would then be reasonable to theorize that the popular "heavens" and "hells" described in many religions and cultures could actually be places in these invisible higher energy and higher dimensional dark matter counterparts of Earth, with their own life-forms and landscapes. They would be invisible to the cognitive-sensory systems associated with our ordinary matter bodies and our current scientific measuring systems, composed of ordinary matter. Furthermore, these heavens and hells would be located at many levels, as dark plasmaspheres are structured into concentric shells around the Earth, as discussed in chapter 8.

Many cultures have situated their gods at the summits of mountains, for e.g., Mount Olympus in Greek mythology, Mount Sinai in Judaism and Christianity, and Mount Kailash in Hinduism. These cultures may have witnessed sightings in the past of gods high up in the atmosphere and within the neighboring dark physical-etheric or astral plasmaspheres, superimposed against the summits. The practice of looking up to the sky to pray (in whatever country around the globe) may be quite correct since a number of dark matter heavens (and their beings) could actually be located within and beyond the Earth's atmosphere. Many cultures have also consistently situated their hells below the surface of the Earth. As discussed in the next chapter, these may actually be situated in the lowest energy Dark Earth, which is closest in energy levels to our ordinary matter Earth.

Spectrum of Dark Earths

Each of these Dark Earths, co-rotating with the ordinary matter Earth, sit inside their own huge bubble universe. Assuming an ascending pathway from our ordinary matter Earth, there would be a spectrum of these dark spheres, which would increase in energy levels, temperature, frequency, entropy, and dimensionality. The number of spatial dimensions will increase until it reaches ten, when the time dimension completely morphs into a spatial dimension. Elementary particle masses and densities will fall while the energy density of each bubble universe's quantum vacuum, and the kinetic energy and frequency of particles, increases. The particles will become progressively warmer, resulting in spheres with larger volumes. Each succeeding universe will become less matter-like, and more radiation-like. Time will become increasingly space-like. From a predominantly classical reality, the body will experience a phasing into a semi-classical reality, and then a quantum reality. Our matter-like reality will then become more mind-like.

Each succeeding and expanding bubble universe (in the ascending energy pathway) will be older (as they were nucleated and created earlier.) This means the earliest universes, specifically from the 10d-particulate photonic sphere, would have already petered out into nothingness due to that universe's accelerated expansion. Therefore, as we ascend to the highest energy levels, the void will increasingly dominate, as fundamental forces become increasingly unified into a single force.

Energy Spectrum of Dark Earths

A series of linked Dark Earths, existing in different universes, which are the counterpart dark planets of the ordinary matter Earth, is presented in ascending order of energy levels (moving towards the high energy vacuum,) below. Each Earth will have a partner – a mirror Dark Earth which interpenetrates and is strongly coupled to it. Each Earth reflects the properties of the relevant bubble universe, brane, or dark sector it inhabits. Only spatial dimensions are indicated. One time dimension should be assumed, except for the 10d universe which has no time dimension.

Energy Level/ Frequency*/ Particle Density	Ref	Dark Earth	Popular Terms in Metaphysics for the Bodies	Relative Reality**
Low Energy/ Low Frequency/ Very High Density	α1	3d-Ordinary-Fluid	Physical-Dense	Classical/ Macroscopic Atomic - Biomolecular Carbon-Based Body in Ordinary Matter sphere
Medium Energy/ Medium Frequency/ Medium Density	α2 β1 β2	3d-Double (Mirror) -Crystalline 4d-Fluid 4d-Double (Mirror) -Crystalline	Physical-Etheric/ Soul Astral-Emotional/ Soul Astral-Mental/Soul	Classical/ Macroscopic Plasma-Electronic Body (Fire-like) in Plasmaspheres

High Energy/ High Frequency/ Low Density	γ1	5d-Fluid	Emotional/ Spirit	
	γ2	5d-Double (Mirror) Crystal	Mental/Spirit	Semi-Classical/ Macroscopic
	δ1	6d-Fluid	Spiritual-Emotional/ Causal/Spirit	Photonic-Plasma Body (Fire and Light-like) in Photonic Plasmaspheres
	δ2	6d-Double (Mirror) - Crystal	Spiritual-Emotional/ Causal/Spirit	
Very High Energy/ Very High Frequency/ Very Low Density	q1	7d-Particulate	Atman/ Monad	Quantum/ Microscopic Elementary Particle-Wave Bodies in Photonic Particulate Spheres
	q2	8d-Particulate	Atman/ Monad	
	q3	9d-Particulate	Atman/ Monad	
	q4	10d-Particulate	Atman/ Monad	

Table 1 - Dark Earths across the Energy Spectrum

* Frequency, among other things, refers to the cognitive frequency of the brain, which is the frequency of transmitted thoughts, and sensory frequency of the body, which is the frequency of received sensory impressions.

** This is the reality that is immediately accessible to the cognitive-sensory apparatus, relative to the relevant body.

Human-Linked Ecological Niches

Just like the ordinary matter Earth, where practically all life-forms live on or just below the crust, plasma, photonic-plasma, and photonic life-forms may not occupy the whole sphere, but certain regions in the plasmaspheres or photonic spheres. Furthermore, the region occupied by these life-forms, which were once coupled or linked with human bodies, would be even smaller.

For e.g., on the ordinary matter Earth, the biosphere basically occupies the crust, which makes up only 1 per cent of the planet. Even on this extremely thin crust, 70 per cent of it is occupied by trillions of aquatic plants and animals who live in the hydrosphere, or the oceans. Within the remaining 30 per cent, there are many large regions that are not conducive to grow large populations of humans. These include places with extremely cold or hot climates – such as Siberia, Antarctica, the Arctic, Greenland, Sahara, and other deserts, and also mountainous regions, such as the Himalayas, Andes, Rockies, and others. According to a World Bank report, 95 per cent of the world's population is concentrated in just 10 per cent of the land surface. Since the land surface is only 30 per cent of the total surface, this means most humans occupy only 3 per cent of the total surface area of the ordinary matter Earth. (Furthermore, since the crust (the land surface) is only 1 per cent of the volume of the Earth, this means humans occupy only a tiny 0.03 per cent of the volume of the ordinary matter Earth.)

While the biosphere on the ordinary matter Earth occupies only the surface of the Earth, the biospheres of even the smallest Dark Earth, for e.g., the physical-etheric Dark Earth, are enormously larger. If there are seven shells in this Dark Earth above the surface of the Earth, it means there are seven biospheres, instead of just one (like the ordinary matter Earth.) Each one of these biosphere would also have a much larger cross-sectional area and volume, because of the longer radius from the center of the Earth, as they are further away from the surface. Additionally, each higher energy Dark Earth has a radius that is five to six times longer than the previous one.

Taking this into account, it would be reasonable to conclude that the *human-related ecological niches* within the biospheres of Earth's dark plasmaspheres can only occupy small and tiny spaces, which became inhabited only in the last ten thousand years, towards the end of the last Ice Age. Based on the nature of structures, the post popular heavens and hells for humans today can be dated to less than a thousand years ago. These historical artefacts include medieval instruments of torture in hells, and metropolises and beautiful cities in heavens. Additionally, UFOs/UAPs in the lowest energy heavens.

The table below identifies the niches that most human-linked plasma life-forms would most likely inhabit in Dark Earths existing in different universes - in ascending order of energy levels, proceeding from left to right, and from bottom to top, from the reader's point of view.

	Dark Earths					
Location on Earth*	3d – $\alpha 1$ (Physical-Ordinary) Plasma	3d – $\alpha 2$ (Physical-Etheric) Plasma	4d - β (Astral) Plasma	5d - γ (Mental) Photonic-plasma	6d - δ (Spiritual) Photonic-plasma	7d, 8d, 9d, 10d- q1, q2, q3, q4 (Formless) Photonic
Not Applicable						Niche 7, 8, 9, 10: Quantum Realities – 4 shells
Beyond Exosphere					Niche 6: Semi-Classical Heavens – 7 shells	
				Niche 5: Semi-Classical Heavens – 9 shells		
Atmosphere, including Exosphere			Niche 4: Classical Heavens – 4 shells			
		Niche 3: Classical Heavens - 2 shells				
Crust	Niche 2A: Classical Human and Animal Sphere – 1 shell	Niche 2B: Classical Etheric Ghosts Sphere – 1 shell	Niche 2C: Classical Astral Ghosts, Demi-gods Sphere – 1 shell			
Higher Mantle		Niche 1: Classical Hells - 4 shells				

Table 2 – Human-Linked Ecological Niches in the Dark Plasma and Photonic Spheres

- *The approximate location is relative to the ordinary matter Earth.
- Each Dark Earth represents a unique planet, and each shell in it represents a unique biosphere.
- Niches 1, 2B, 2C, 3 and 4 are in the imaginal realm, i.e., the classical spheres.

The particle density reduces progressively from the niches in the left to the ones in the right, and from bottom to the top, transforming the reality from a classical one with macroscopic bodies to a quantum one inhabited by elementary particles with consciousness. The energy levels and dimensionality increases in the same direction.

Subjectively, advanced meditators and near-death-experiencers (NDEers), whose locus of consciousness is transferred to classical dark plasmaspheres, will experience being in a macroscopic and personal dark plasma body. In semi-classical photonic plasmaspheres, they will not experience any macroscopic and personal photonic-plasma body until they focus on it. In photonic particulate spheres, they will not experience any macroscopic personal body.

Due to the vastness of the plasma, photonic-plasma and photonic spheres, the above list of human-linked niches is not intended to be exhaustive but only illustrative.

CHAPTER 14

❈

Classical Hells

Earth's Dark 3d-Double Plasmasphere

Before we discuss the classical hells, we will need to understand Earth's dark 3d physical-etheric plasmasphere in which they are located. Carl Jung, the renowned Swiss psychiatrist, and psychoanalyst who founded analytical psychology, recorded a near-death experience (NDE) in his physical-etheric body. He recounted after the experience, "I discovered how high in space one would have to be to have so extensive a view — approximately *a thousand miles*!" 1,610 km (or 1,000 mi) is well above the Earth's thermosphere and is at the lowest part of the exosphere.

A first approximation of its size can be made by working backwards from Leadbeater's estimate of where the larger 4d astral plasmasphere ends. According to his estimate, it ends around the mean orbit of the Moon, which is about 385 thousand km (240 thousand mi) from the Earth. If we take the ratio of the percentages of dark matter and ordinary matter in the universe – i.e., by dividing 85 per cent over 15 per cent, this gives a factor of about 5.5. If we divide 385 thousand (240 thousand mi) by 5.5, we obtain 70 thousand km (44 thousand mi) as the radius of the physical-etheric Dark Earth.

Interestingly, Stephen Adler of the Institute for Advanced Study (Princeton, NJ) estimated in 2009 that about twenty-four trillion metric tons of dark matter lie between the Moon and Earth and that the dark matter density above the surface of the Earth peaked at about 70 thousand km (44 thousand mi) from the center of the Earth. This dense region, located in the exosphere, and functioning as a natural electrode (in this case a cathode,) would most likely be sheathed (with a double-layer) to ensure the overall charge-neutrality of the dark plasmasphere.

In terms of volume, calculated from the radius, this dark plasmasphere will be three orders of magnitude or one thousand times larger in volume than the rocky Earth. Hence, the physical-etheric 3d-double would be a Jupiter-sized dark plasmasphere, within which the rocky Earth sits. This means it can host trillions of life-forms, with its numerous shells representing huge biospheres, each one being much larger than the ordinary matter Earth's.

This Dark Earth sits inside the physical-etheric universe which interpenetrates the ordinary matter universe. As discussed in chapter 11, it is a slightly higher-energy CPSTM-reversed mirror universe. (CPSTM: Charge, Parity, Space, Time, Matter/Force symmetries.) It could also be considered a shadow universe, being one of the E6 components of the "E6 x E6" superstring model of the universe. This Dark Earth can be considered a planet (in its own right) which is gravitationally-coupled to our ordinary matter Earth and possesses a rotational speed and gravitational force that is different from the ordinary matter Earth. Just like the ordinary matter Earth's plasmasphere, this plasmasphere would have a "tear-drop" shape, three spatial dimensions and one time dimension.

Signature Features of Plasma in Dark Earth

Like the other dark plasmaspheres that will be discussed, Earth's 3d-double plasmasphere would contain signature features of plasma – including concentric shells, filamentary currents, and plasma vortexes.

Filamentary Currents

Filamentary currents are present not only in space but in our plasma bodies and also enveloping the dark plasmasphere. These currents in Dark Earth have been referred to as "ley lines" in the West or "dragon lines" in Chinese Feng-Shui studies. They are analogous to the "meridians" identified in Chinese acupuncture or "nadis" in Indian yogic literature. They include both currents within double layers, and "Birkeland currents," tracing magnetic field lines, that often cut across it (as discussed in chapter 7.) It is not surprising, therefore, that Chinese acupuncture describes the various meridians in subtle bodies as components of "microcosmic-orbits" or circuits. Birkeland currents imply *electric circuits*, which follow Kirchhoff's circuit laws.

Plasma Vortexes

In plasma physics, we note that when filamentary currents cross, they pinch, collapsing to form nodes, which give rise to vortexes. Similarly, when ley or dragon lines, or meridians cross, they generate vortexes. In Hindu metaphysics, these vortexes are described as chakras (which simply means "wheels" in Sanskrit.) Depending on the direction of spin, vortexes can absorb or emit energy. They have been identified as "holy" or "sacred" sites on Earth by various spiritual traditions. Ley lines, or leys, connect ancient holy sites — just as filaments connect vortexes.

Higher Frequency Dark Light in Earth's Twin

As discussed in chapter 8, the dark physical-etheric universe has its own light, which we can call "dark light." This light is much more energetic and hence at much higher frequencies than ordinary light. This provides much higher resolution to the life-form's vision than the light in the ordinary matter universe. Hence, near-death-experiencers (NDEers) will often see and report much clearer and vivid hyper-real views of the environment, while they are operating from the dark etheric double, compared to what they see in their ordinary vision,

Now that we have some understanding of the environment of the 3d Dark Earth, we will explore the ecological niches occupied by human-linked plasma life-forms, with a focus on Niche 1, or the classical hells in the lowest energy dark plasmasphere.

Classical Hells

Classical heavens and hells can be explained with classical physics. This physics relates mainly to the behavior of macroscopic objects and the averaged behavior of large collections of elementary particles. It includes Newton's laws of motion and gravity, Einstein's theories of relativity, and thermodynamics.

On Release from the Ordinary Matter Bodies

When the physical-etheric body is discarded at the same time as the ordinary matter body or immediately after, and the astral body is heavy, it will remain in Niche 2C in the astral plasmasphere, coincident with the Earth's surface – and not rise up to the lowest astral heaven in Niche 4. The discarded physical-etheric body or parts of it, in Niche 2B, may still retain some reflexes of a life-form, and may be encountered as ghosts resembling unintelligent automatons.

When the physical-etheric body is not immediately discarded, after it is freed from the dead ordinary matter body, it will be in Niche 2B in the physical-etheric plasmasphere (which is coincident with the surface of the ordinary matter Earth.) Now that the physical-etheric ovoid is separated from the ordinary matter body, the low-energy particles will no longer be confined and compactified within the humanoid body within the ovoid. It will disperse into the plasma in the ovoid. The centrifugal force in the rotating concentric shells within the ovoid will push a large number of these particles to the outermost shell. This blocks-off communications with the physical-etheric shell in Niche 2B, which is coincident with the surface of the Earth (where family and friends are.) This would cause a black-out which may be subjectively construed as a loss of vision.

However, it takes some time (about a week) for the whole process to work its way through. During this time, the physical-etheric plasma-based being (popularly called a "soul" or "spirit") can visit friends and relatives, usually with a personal guide. This guide is a plasma-based being who has been assigned to the person and is sometimes referred to as a "guardian angel" of that person. Despite the term "angel," which conjures up images of humans with wings, this guide will have an appearance that conforms to the belief system of the person exiting the ordinary matter sphere. The appearance is generated through plasma dynamics, as explained in chapter 10. These guides are basically doing their jobs and may have more than one person to look after. They are not "personal friends." Sometimes, they can even be quite temperamental. They will not appear if they see that you do not need any help and are going in the right direction.

As the physical-etheric body has been coupled with the ordinary matter body for a lifetime, the dark etheric matter has become highly compactified and consequently dense (as discussed in chapter 9.) If it is not too heavy, after the "black-out" or loss of vision sets in, most people will rise spontaneously or be escorted by a guide into the lowest energy heaven in Niche 3, between the troposphere and stratosphere, about 20 km above sea-level. If it is very dense and heavy, after the "black-out" or loss of vision sets in, the heavy and compactified physical-etheric body quickly sinks or gravitates into what is popularly called "hell," together with other physical-etheric matter of the same density, as a natural consequence of its specific gravity.

This region will be in Niche 1, which is coincident with the uppermost layer of the hyper-viscous mantle, just below the surface or crust of the Earth (see Table 2 in chapter 13.) Most religions situate their hells here (see chapter 2.) This is expected as this region contains a high density of some of the lowest-energy dark matter particles, near the core of the lowest energy physical-etheric plasmasphere. Here vision will be restored and (undesirable) communications become possible as the lower-energy particles that lined the outer shells of the ovoid can now respond to the environment, which has similar properties. Jains believe that karma is physical and composed of particles that can be found everywhere in the cosmos, which alludes to dark matter particles. Dense, low-energy, dark plasma can be equated to heavy karma.

> *Heavy* astral karma must be redeemed by such beings before they can achieve an unbroken stay in the causal thought-world (emphasis added.)
>
> Paramahansa Yogananda, Mystic

Unpleasant Region

The pressure imposed by the dark physical-etheric matter, in the huge physical-etheric sphere above the region, is enormous, causing the matter to be compactified and compressed in Niche 1. Although the pressure is great by terrestrial standards, it is relatively mild when compared to pressures within stars, so it will not cause nuclear fusion or superfluids (or quantal fluids) to be formed. Based on density correlations between dark and ordinary matter, the pressure in the core of the physical-etheric plasmasphere will be expected to be relatively higher than the rest of the sphere, just as in the ordinary matter Earth.

> The pressure at the center of the Earth is around 14.2 million times [of the] atmospheric pressure. Not a nice place for a visit, but quite mild compared to the center of a star.
>
> Nich Hoffman, Geologist

Due to the high density, this niche is indescribably loathsome to the liberated physical-etheric body - movement is restricted and the environment is claustrophobic. Moving through such an environment (in the dark etheric body,) according to metaphysicist Leadbeater, would be like pushing your way through a black, hyper-viscous fluid. As expected, the density in this dark plasmasphere correlates to the hyper-viscous liquid mantle in the ordinary matter Earth. The deeper it goes, the more compressed, compactified and dense the dark plasma body becomes.

> The negative visionary finds himself associated with a body that seems to grow progressively more dense, more tightly packed...many of the punishments described in the various accounts of hell are punishments of pressure and constriction.
>
> Aldous Huxley, Author of "Heaven and Hell"

If the individual was in a dream state, the physical constriction may conjure up a myriad of imaginative dream scenarios. Research on the incorporation of sensory stimuli into dreams show that both pressure and temperature can influence the content of a dream. In a study conducted by T. A. Nielsen in 1993, participants wore a cuff on their leg while sleeping in the laboratory, which was inflated during REM sleep to cause pressure. In the dream reports on waking, the researchers found several examples of leg pressure incorporated into dreams.

Hellish Dreamscapes

Certain religious communities, in particular the Hindus, Buddhists, post-first century Christians, Catholics and Muslims, have created their own specific hells corresponding to concepts in their religions and societies. These are called Belief System Territories (BSTs) by experimental metaphysicist Robert Monroe. These hellish BSTs (there are also heavenly ones) are in the "imaginal realm," as described by Kenneth Ring, Professor Emeritus of psychology at the University of Connecticut. This realm is in between imaginary dreams and matter-like reality, between mind and matter. The term was coined by Henry Corbin, a theologian and Sufi mystic.

While normal dreams are generated by individuals, these are by the collective unconscious. As a result, they would have a quasi-independent existence, outside the control of an individual. They would feel hyper-real and much more lucid than normal dreams, due to the higher resolution of dark light. The hellish content reported in NDEs therefore goes beyond subjective feelings and is objectively frightening. It describes landscapes, entities, or sensations that are unworldly and frightening. (Please see chapter 10 on the technicalities relating to the formation of dreamscapes in a plasma environment.) The fact that they are shared dreams is evident from numerous observations.

Firstly, the recorded observations usually recount instruments of torture that are archaic and medieval, which are not relevant today. Secondly, the individual can be cut or burned, but still manages to resurrect and be subject to the same torture over and over again. Plasma life-forms have finite life spans. Thirdly, dark plasma bodies are not ordinary matter flesh and bones – they are unlikely to be harmed by heat (in fact, they would gain energy) and they can't be "cut" by medieval saws which will just pass through it.

As can be expected, the low-energy plasma life-forms native to this region, would have a diminished state of consciousness and narrow perspectives, so they would generally be malevolent. They may opportunistically use the dreamscapes to terrorize transient human-linked plasma life-forms.

Hells are Created by Living People

These brutal hells are constructed on a daily basis in the imaginal realms of our nearest neighboring Dark Earth with trillions of thought-modulated dark electromagnetic waves being generated by the growing population of human beings on Earth, and previous societies (in total, almost 117 billion people.) It is quite evident from the descriptions of the most horrifying hells, that they are predominantly from the minds of obsessively vindictive people.

Many religions created these etheric concentration camps in the imaginal realm through the repetition of teachings relating to bizarre methods of torture, improvising as they go along, to boost the number of converts instantly. The more gruesome, the quicker the conversions. However, these gains are short-term. They will soon realize that, in the very hells that they are creating in the imaginal realm, are their mothers, fathers, brothers, sisters and beloved friends who will be terrified and tortured psychologically. The modern preachers who continue to do this are actively creating hells for themselves and their loved ones. Although built-up by a large number of separate individuals with similar thoughts, when the thought-holograms coalesce, they create hells that are difficult to control and that have a life of their own.

To eliminate this vicious cycle of obsessive and compulsive vindictiveness, it is necessary to forgive everyone, particularly your enemies. This will reduce the build-up of vengeful thoughts that find expression in the excessive and senseless punishments in these imaginal hells. You can go one step further by not having a mindset that creates enemies in the first place, by understanding and empathizing with people who have different views. Hopefully, as society becomes more informed by science, with more people rejecting religious superstitions and becoming more serious about human rights and treating people humanely and compassionately, these hellish dreamscapes will disappear. And then, we will see it for what it was meant to be – a natural detoxification process.

The Ultimate Sauna

What is actually happening in the background, outside the dream, is that due to the higher temperature, particles in the dark plasma body are gaining enough kinetic energy to increase the volume of the body and also expel heavier low-energy particles. This will reduce the body's density (reversing the effects of any compactification) – so that it begins to float up and out of the shell to the surface of the Earth. From the perspective of this book, hell is therefore a place for detoxification, just like any sauna (or volcanic spring,) to remove toxins from the recently arrived dark plasma body, acquired from living in the ordinary matter Earth, which may not be conducive to living in the next destination, most probably in some shell in the dark physical-etheric or astral plasmaspheres. The purification not only detoxifies the physical-etheric body, but also the astral and other higher energy bodies, as a side-effect.

The descent to hell is a natural phenomenon and is temporary, although the actual duration can vary significantly for different persons, depending on the amount of toxins. More religions are embracing this concept of detoxification or purification. For e.g., the concept of purgatory, introduced by Catholics in the eleventh century AD is a step in the right direction. Purgatory is a place of purification prior to entering heaven. Going forward, religious teachers and preachers should emphasize on restorative, rather than an infantile retributive justice.

Hells are not Eternal in a Love-Based Framework

It is interesting to note that out of the thirty-one planes identified in Buddhist cosmology, only one relates to the hell realm (which contains, within it, different levels of hells.) This is only about 3 per cent of the total. The hellish planes also occupy less space as they are mostly in the lower volume plasmaspheres (i.e., the physical-etheric sphere,) in a region enclosed within a smaller radius. So, we would expect less exposure to hells than heavens for the journeying dark plasma bodies. This is supported by NDE (near-death experiences) research. Less than 2 per cent of all NDEs shared with the NDERF (NDE Research Foundation) are hellish.

According to this research, most stressful NDEs are not frightening throughout the experience but also contain pleasant parts. When both are present, the frightening part is usually shorter and occurs at the beginning, followed by the longer pleasant part. Heavy physical-etheric bodies will need to descend into hell first for a short time, to detoxify, before ascending into, usually, the lowest heaven (as discussed in the next chapter.) According to numerous recorded cases, it is extremely rare for a pleasant NDE to turn into a distressing one. Those with hellish experiences were more likely to be less alert during their NDE, than those who have pleasant NDEs. This suggests that they were in a drowsy dream state and not fully conscious of their actual environment, making them more susceptible to shared collective dreams.

Some who experienced distressing NDEs say that when they "gave up" fighting the NDE and surrendered to it, or sincerely asked for help from a loving "Higher Power," their distressing NDE changed to a pleasant one. As noted in chapter 9, stress has the effect of compactifying the dark plasma body to make it more dense and heavy. The more distressed the person becomes, the more he/she will sink. On the other hand, a relaxed mental state will reduce the density of the body and allow it to float up. This may be one of the most straight-forward ways of avoiding sinking into hell during the death process. The process is very much like floating in water, which is well captured in the lyrics of John Lennon in the Beatles' song, *Tomorrow Never Knows*: "Turn off your mind, relax and float downstream. It is not dying; it is not dying." However, this will be difficult for those who have extreme attachments or aversions, as well as those who have other unfinished business here. They will not be able to relax due to their agitated states, which will compactify their bodies further.

Extremely evil people would therefore be subject to longer durations of hellish dreams. Even before this, they will undergo a life review where they will relive all the terror of their victims in minute detail. This will increase empathy for the victims. They will also be reflecting on their behaviour from the vantage point of a higher state of consciousness (with a larger cognitive capacity) and so will realize the horrors that they have inflicted more clearly. This constructive approach in the NDE is based on restorative justice, which encourages reconciliation and universal love. It is also fair, as perpetrators go through exactly what the victims did. It is not the retributive and punitive justice that traditional religions have been promoting to get more converts.

Back to Classical Earthly Worlds

With the high temperatures in this region, the particle density falls over time, and the bodies become lighter. The physical-etheric plasma bodies, which are trapped higher up in the shell, would then be able to escape and rise to Niche 2B (coincident with the surface of the ordinary matter Earth,) where, with the new composition of the plasma body, communications now become possible with former friends and family.

Most people tend to hover around their earthly homes, in order to stay connected with their friends and the places which they know for a brief time. Subsequently, their physical-etheric bodies die, and they phase into the astral plasmasphere. The majority of beings at this stage spend some time comparatively near to the surface of the ordinary matter Earth, within the 4d (astral) plasmasphere. This is in Niche 2C of Earth's dark 4d astral plasmasphere (see Table 2 in chapter 13.) Note that there are many different pathways between the niches, depending on specific conditions. The above discussion is only about the most probable pathways.

CHAPTER 15

❈

Classical Heavens

Earth's Dark 4d Astral Plasmasphere

> There is no reason why a four-dimensional world could not exist...or for that matter a world of five dimensions or six or seven.
>
> Martin Gardner, Popular Science Writer

The dark astral universe is a "hypersphere" with four space dimensions and one time dimension. However, an observer, who is using a dark plasma astral body, linked to an ordinary matter body inhabiting a 3d universe, will only see the 3d-correlate for some time. Earth's dark 4d astral plasmasphere inhabits this universe. It exists all around us and interpenetrates the ordinary matter Earth. Most human beings would find themselves in this 4d astral plasmasphere after the death of their ordinary matter and physical-etheric bodies.

It can be considered a planet, in its own right, having its own rotational speed and gravitational field. Just like the ordinary matter Earth's plasmasphere, the dark astral plasmasphere would generally conform to a tear-drop shape, while shape and size-shifting like a candle flame. This is due to gravitational interactions with the Sun and Moon, as well as dark plasma winds from the Sun and the Milky Way.

Dimensions of the Dark Astral Plasmasphere

The astral Dark Earth extends beyond Earth's exosphere. According to Leadbeater, it extends to a little less than the average distance of the Moon to the Earth, which is approximately 385 thousand km (240 thousand mi.) The Earth's diameter is not more than 13 thousand km. The average distance between the Earth and Moon is, therefore, approximately thirty Earth-diameters. Based on these measurements, it can be calculated that the 3d-correlate of the 4d astral plasmasphere would be more than 200 thousand times larger in volume than the ordinary matter Earth. This dark or shadow planet is therefore much larger than even Jupiter. (The strongly coupled dark plasmasphere in the mirror universe will be about the same size.)

If you were holding a basketball on the tip of your index finger; the astral world would represent the basketball and the tip of your index finger, the ordinary matter Earth. Within this massive sphere are located all our more popular classical heavens, mostly above large concentrations of the corresponding religious communities. Unlike the ordinary matter Earth, where the biosphere is confined mainly to the very thin crust, there are numerous biospheres in each of the various shells in the dark astral plasmasphere, with each one being much larger. It would contain trillions of plasma life-forms, including a small population of human-linked plasma beings in the relevant ecological niches.

Rotation of the 4d Plasmasphere

Being a much larger sphere, the rotational speed of the dark astral plasmasphere would be slower, making a day in the astral plasmasphere longer than 24 hours. The mystic Yogananda has confirmed that the astral day and night are longer than in the ordinary matter Earth. Since astral objects emit light, even the nights would be filled with a diffused light in the astral plasmasphere. All this may give a misleading impression to some that there is no day and night, when, in fact, there is. Based on plasma dynamics, counter-rotations of some shells in the plasmasphere cannot be ruled out. The astral year would be as long as the ordinary year, but the number of days would be less.

The actual locations of persons (who have left their ordinary matter bodies permanently) can be identified approximately in the classical heavens that populate the astral plasmasphere, based on his/her belief system and the density of his/her dark plasma body. A predominantly Catholic country would have a heavily populated Catholic heaven several kilometers above its surface but at a different longitude. Similarly, a predominantly Hindu country would have a heavily populated Vedic-influenced heaven several kilometers above its surface, but with some difference in the longitude. There is a longitude and time displacement due to the different rotational speeds of the ordinary matter Earth and its astral plasmasphere. The individual can therefore be located in the relevant Belief System's Territory, after a certain period of detoxification in the physical-etheric plasmasphere. However, over time, their equilibrium positions in the 4d (astral) plasmasphere changes as the properties of their plasma body, as well as their beliefs, change. They would then be more difficult to locate.

Advanced meditators, who have access to these heavens (as well as the hells in Niche 1,) often visit the ones closest to their location in the ordinary matter Earth. As such, they may visit the heavens and hells created by their belief systems, reinforcing their beliefs in their systems. On this basis, they may develop a confirmation bias that leads them to reject other belief systems. In fact, most of these long-standing belief systems, with large followings, have their own heavens and hells. However, no single religion has a monopoly over all the heavens and hells.

The Moon and the Earth

Just like the Earth, the Moon has its dark matter counterparts – the 3d-double, 4d, 5d and 6d dark plasmaspheres. When the Moon comes closer to Earth in its orbit, there is an overlap between the highest shells in the 4d (astral) plasmaspheres of the Earth and the Moon. When the plasmaspheres of the Moon and the Earth merge, higher energy beings from both astronomical bodies can travel back and forth and mingle. Plutarch, the Greek philosopher who lived in the first century AD, wrote, "Every soul is ordained to wander between incarnations in the region between the Moon and Earth." (Hence, the title of this book, "Between the Moon and Earth.")

It is reasonable to assume that the physical-etheric, astral, and higher energy plasmaspheres of the Moon and Earth would mutually gravitate towards the center of mass of the Earth-Moon system — causing tides in Earth's plasmaspheres – which are dynamically similar to oceans. There would be an "ebb and flow" of etheric, astral, and higher energy dark matter. Our dark plasma bodies would feel this, causing changes in our emotional and mental states through subtle interactions with biochemical fields in our ordinary matter body.

Communities in the Astral Plasmasphere

According to Paramahansa Yogananda, the recently physically disembodied being arriving initially in the astral plasmasphere, joins an astral family through invitation, drawn by similar mental and spiritual tendencies. The Swedish mystic, Swedenborg, confirms this:

> People who arrive from the physical world are put in connection either with a particular community in heaven or with a particular community in hell; but this applies only to their more inward elements.
>
> Emanuel Swedenborg, Mystic

From the perspective of physics, we would explain that an astral body will gravitate or levitate to shells that are composed of plasma that have similar properties to the plasma in their bodies, which is a consequence of the lifestyle and mental states of the owner. Due to plasma dynamics, individuals with similar characters will therefore tend to congregate in the same shells. Guides from the plasmasphere will assist in this process if they think it is required. Subsequently, the individuals in the astral Niche 2C (coincident with the surface of the ordinary matter Earth) grow in awareness. They then withdraw into themselves, and their locus of awareness shifts to higher energy shells farther up. They will then find it more natural to soar farther away from the ordinary matter Earth's surface into Niche 4.

Classical Heavens

Niches 3 and 4 contain several classical heavens which are at various altitudes in Earth's atmosphere (including the exosphere.) Most of the heavenly experiences during near-death experiences occur in Niche 3 in the physical-etheric plasmasphere. If they are still linked to their living ordinary matter bodies, humans will not be able to cross over permanently to the astral plasmasphere. Buddhists call these regions the "Sensuous Heavens" – of which there are six shells (2 in the physical-etheric plasmasphere and 4 in in the astral plasmasphere.) They are called "sensuous" to emphasize that there are sensory impressions that are quite similar in nature to the visual, audio, and other impressions while in the ordinary matter body – although these are at much higher frequencies and energy levels. We will call it the "Classical Heavens," which captures the actual meaning more accurately from a scientific perspective. It is also part of the imaginal realm – the realm between mind and matter created by large groups of conscious agents.

As one example of the type of heavens here is the lowest of these heavens in the physical-etheric plasmasphere, as recorded in Buddhist cosmography, the "Yama" heaven. It is closest to the ordinary matter Earth's surface, and would be coincident with the upper troposphere, and lower stratosphere. This is only about twenty km above sea-level, but well above Mount Everest and the flightpaths of most commercial aircraft. The heaven just above that is called the "Tusita" heaven, where Bodhisattvas use as a transit location, before they return to the ordinary matter Earth to complete their missions. The historical "Buddha," is thought to have descended from this heaven as a Bodhisattva into Niche 2A (the human and animal ecological niche in the ordinary matter Earth.) This heaven would be coincident to the middle stratosphere – about forty km above sea-level. The heavens discussed above are only examples for illustrative purposes. There are many other heavens belonging to different Belief System Territories (BSTs), all over the globe.

The Gospels recount that Jesus ascended into heaven, after forty days on the surface of the ordinary matter Earth. As he ascended, his physical-etheric body (which would look very similar to his ordinary matter body and include any recent high impact wounds) entered a classical heaven. The most proximate and probable one would be a heaven that is at the same altitude as the classical Tusita heaven. However, while that was above India, this would be above the Mediterranean Sea. This is in the stratosphere, and west of Israel, due to the slower rotating astral plasmasphere (as discussed above.) Hence, the travel route would be upwards, and then westwards. The travel trajectory would be similar to most human-linked plasma bodies, except that most former humans would get off at the first "station," at the altitude of the Yama heaven, about twenty km above sea level. The Tusita heaven or equivalent, that Bodhisattvas go to, is the second station, about forty km above sea level.

> In the beginning, it is that of climbing higher and higher and afterwards it usually takes a westerly direction.
>
> Hiralal Kaji, channeling a spirit's account of his journey

Dreamscapes Created by Religious Beliefs

As noted in the previous chapter 14 for classical hells, but this time with respect to classical heavens, these are regions that experimental metaphysicist, Robert Monroe, would call "Belief System Territories (BSTs)." They are also dreamscapes, like the classical hells, with hyper-real and lucid appearances, due to the higher frequency of dark light. The beings here collectively mould their surroundings with their more persistent thoughts.

According to Monroe, in the heavenly BSTs we will find groups of people, with shared religious beliefs when in the ordinary matter body, congregating. There will be many heavens, and hence BSTs, of all the major religions which have been generated by religious communities over centuries. People who arrive here live according to what has been taught to them during life, as to what to expect in heaven. They live in imaginary cities of their own, partly creating them entirely of their own thoughts, and partly inheriting and adding to the structures created by their predecessors, very often corresponding to the scriptures they have read. The structures are objectively real to all the inhabitants.

According to Michael Talbot, near-death-experiencers (NDEers) and others describe great cities that are brilliantly luminous, with remarkable consistency. The Swedish mystic, Swedenborg, said that it was a place of "staggering architectural design, so beautiful that you would say this is the home and the source of art itself." These sensuous classical heavens are extremely pleasurable and beautiful. As a result, earthly associations become less important for the plasma bodies that have soared to Niches 3 and 4 (from Niche 2B or 2C on the surface.)

A person will reside in one of these, sometimes also described as "hollow heavens," for a long time before noticing inconsistencies in the religious doctrines. When they can no longer resonate with those present in that particular hollow heaven, they will move to a higher energy shell of classical heavens, higher up in the atmosphere, or beyond it.

In addition to human-linked plasma life-forms, there are also billions of native plasma life-forms in the various shells of the plasmasphere, that outnumber them by several multiples. (This should not be surprising as even on the ordinary matter Earth, insects are the most common life-forms, followed by fishes, amphibians, reptiles, birds, and mammals. Humans make up a minute proportion.) Some of the most common plasma life-forms in this plasmasphere have been characterised by humans as gods, deities, angels, and other similar entities.

Conclusion

As discussed in chapter 11, the lower frequency 4d astral plasmasphere has a propensity to manifest and objectify emotions and thoughts that are in lower frequencies, while the 5d mental (or intellectual) plasmasphere, in higher frequencies (as discussed in the next chapter.) This suggests that dark matter particles in the 4d plasmasphere are heavier and less energetic than the 5d plasmasphere. We can surmise that dark matter particles in each succeeding higher energy plasmasphere or dark sector are lighter and more energetic as they correlate to more subtle states of consciousness. Emotions, thoughts, and the experience of joy and peace would therefore become increasingly subtle as we ascend to higher energy plasmaspheres. In the next chapter, we will discuss the even larger 5d and 6d dark plasmaspheres, where energy levels and frequencies are much higher.

CHAPTER 16

❈

Semi-Classical Heavens

> Beyond the heaven visited, there was another heaven, one so brilliant and formless to perception that it appeared only as a "streaming of light."
>
> Saying attributed to Emanuel Swedenborg, Mystic

The human-linked ecological niches of Earth's 5d and 6d plasmaspheres can be described as semi-classical heavens. These are intermediate between classical heavens (covered in the previous chapter) and quantum realities in particulate photonic heavens (that will be discussed in the next chapter.) They require both an understanding of classical, as well as quantum physics.

Quantum physics studies the behavior of elementary particles at microscopic levels. Its laws do not function the same way as macroscopic objects, which are more efficiently described by classical physics. While classical heavens, hells, and dark plasma bodies, can be largely explained using classical magnetohydrodynamics (MHD), kinetic theory, electrodynamics and gravity, semi-classical heavens would also require quantum hydrodynamics and superfluid MHD. Quantum hydrodynamics is the study of hydrodynamic-like (i.e., liquid-like) systems which demonstrate quantum mechanical behavior. Superfluid MHD relates to the MHD of plasma superfluids.

Location and Size

The semi-classical heavens occupy ecological niches in the 5d and 6d dark plasmaspheres. Leadbeater estimates the size of the 3d-correlates of these higher energy Dark Earths to be the same ratio as the astral to the physical-ordinary sphere (which is the same as the physical-etheric plasmasphere.) For the 5d plasmasphere, this could be computed by taking the factor of 5.5 (the ratio between the physical-etheric and the astral plasmasphere) and multiplying it by the size of the astral plasmasphere which extends to about 385 thousand km (24o thousand mi) from the center of the Earth. This gives 2.1 million km (1.3 million mi.) The 3d-correlate of Earth's 5d plasmasphere would therefore extend beyond the orbit of the Moon and will overlap with its dark plasmaspheres. (The strongly coupled dark plasmasphere in the mirror universe will be about the same size.) Computed similarly, the 3d-correlate of the 6d plasmasphere will be about 11.6 million km (7.2 million mi.) (The strongly coupled dark plasmasphere in the mirror universe will be about the same size.) This is well beyond the Moon's orbit, but significantly closer than the closest planet, Venus, which is approximately 40 million km (25 million mi) away from Earth. Mars is much further away.

Like the other plasmaspheres, these also have a tear-drop shape, and interpenetrate all the lower-energy Dark Earths and the ordinary matter Earth and are gravitationally bound to each other. Each of these Dark Earths could be considered a dark or shadow planet in its own right, having its own length of day and night, which would be different from ordinary matter Earth's due to the difference in rotational speeds. Some researchers are puzzled as to where the "planets" in Vedic cosmology are located. They are, in fact, these ordinarily invisible Dark Earths. The distances of the various levels of heavens in Vedic cosmology seem to be largely exaggerated. Nevertheless, they do approximate the estimates above.

Nature of Semi-Classical Universes and Bodies

The composition and behavior of higher energy plasma bodies in these plasmaspheres, and their universes, are based more on "photonics," rather than "electronics," compared to lower energy spheres – more "light-like," and "radiation-like," rather than "matter-like" (as discussed in chapter 9.) Swedenborg put it quite succinctly when he said that these heavens were like the "streaming of light," i.e., streams or rivers of liquid light. Yet, it was also formless. Hence, they are in between the gross forms of the classical realms and the formless realms in the particulate photonic spheres, having attributes of both. Time is almost absent.

The dark plasma bodies here are composed of a much higher proportion of highly energetic neutral dark photons, mixed in with dark electrons and protons. Since complex plasma is a liquid-crystal, these plasma-based "light bodies" will display characteristics of both light-crystals and liquid-light. Photons, which are particles of light, do not have a charge and they do not feel the electromagnetic force. However, light is, in essence, oscillating electric and magnetic fields, as it is an electromagnetic wave. Charged particles will therefore experience forces when they are in these fields. Due to the extremely low mass and velocity of these photonic-plasma bodies, the wave-like nature of their bodies will dominate due to the de Broglie relation in quantum physics.

When not being measured or observed, by the owner or others, they will not exist as definite bodies but be smeared out in spacetime. Subjectively, advanced meditators and near-death-experiencers (NDEers), whose locus of consciousness is transferred here, will sense a new radiant universe but not a personal body until they make a conscious attempt to observe their own body. It will then spontaneously materialize, in the same sense that a wave materializes as an elementary particle when observed or measured. Buddhists might describe this as a "spontaneous birth." This is correct from the classical perspective. However, we now have deeper insights from quantum physics.

Dark Photonic-Plasma Bodies

Massless photons can bond to form photonic molecules so strongly that they act as though they have mass. In plasma, when photonic molecules gain mass, they can form structured hybrid photonic-plasma life-forms. Scientific researchers also theorize that Bose-Einstein condensations (the fifth state of matter) of photons can occur within high temperature plasma (the fourth state of matter,) notwithstanding that these normally occur at extremely low temperatures. These condensations behave quantum-mechanically even at macroscopic scales. Within semi-classical macroscopic dark photonic-plasma bodies in these photonic plasmaspheres, they give rise to both classical and quantum-mechanical properties and behavior. Gender characteristics are absent in these androgynous plasma-based light bodies.

Cognitive-Sensory Systems

Cognitive capacity increases significantly in these life-forms due to a higher dimensional neural structure and quantum processing using photons. They use electromagnetic wave (instead of acoustic wave) cymatics to generate thought-holograms. Lasers issue from vortexes for various purposes, including healing. The dark matter in these semi-classical dark photonic plasmaspheres vibrates in response to more subtle mental activities, abstract thoughts, and emotions, compared to the matter in classical dark plasmaspheres. The life-forms here use photonic sensors, and their sensory perception up-shifts to higher frequency optical and acoustic sensations, beyond lower frequency smell, taste, or touch, and with an expanded bandwidth. All this results in a state of super-consciousness.

Communications and Transport

Dark electromagnetic waves, thought-holograms and bubble-drives are used for communications. For transport and locomotion, photonic propulsion with lasers are also used in these extremely subtle universes. (The details of these mechanisms have already been discussed in chapter 10.)

Semi-Classical Ecological Niches

Niche 5 – Photonic Plasmasphere

Niche 5 is in a five-dimensional plasmasphere that is much less dense than the 4d plasmasphere. While other religions would have their own Belief System Territories (BSTs), Buddhists and Hindus would identify these as the lower 9 shells of "Brahmaloka." These spheres attract very advanced meditators, sages, and abstract thinkers, who live here for very long periods of time. The life-forms in this sphere possess macroscopic hybrid photonic-plasma bodies.

Niche 6 – Higher Energy Photonic Plasmasphere

Niche 6 is in the six-dimensional plasmasphere, with a much higher energy level than the previous niche. While other religions would have their own Belief System Territories (BSTs), Buddhists and Hindus would identify these as the higher 7 shells of "Brahmaloka." The most powerful gods live in these spheres for very long periods of time. The life-forms in this sphere possess macroscopic photonic-plasma bodies, which become increasingly photonic light bodies. In the highest shell, it takes the form of a photonic crystal.

Turning Point

The 6d universe represents a significant turning point, from the classical and semi-classical realities of macroscopic life-forms in plasmaspheres and large astronomical bodies in plasma universes, to quantum realities in the near-void photonic particulate spheres (to be discussed in the next chapter.) It represents the limit of the operation of the E6xE6 symmetry. It is usually called the "causal world" in the wider metaphysical literature, being the first sphere in which a large-scale creation of macroscopic bodies takes place. (Strictly speaking, it is the 10d particulate sphere (see next chapter) which is the causal sphere – i.e., the sphere where time and causality (including what religions would loosely call karma) first becomes weakly operative.)

CHAPTER 17

❋

Quantum Realities

In the highest energy spheres (i.e., the 7d, 8d, 9d and 10d spheres,) particle densities are incredibly low. Hence, there are no macroscopic bodies or plasmaspheres. Only microscopic "particle-bodies," which are widely separated in a higher-dimensional quantum vacuum, exist. Hence, they are called "particulate spheres." Here, dark matter particles are non-interacting (in a classical framework) and neutral. The author conjectures that they are in the nature of highly energetic and massless dark *photons*. As the speed of light and causality falls in higher energy universes, they will be very slow-moving. This accords well with astrophysical observations which show that the bulk of neutral dark matter particles move very slowly.

These highly energetic dark photons will collectively warp spacetime, generating shadow gravity that can be measured in our ordinary matter universe. Due to the large inter-particle distances, the photons become isolated and disentangled from the environment. As quantum objects, they have particle-like properties when being measured, and wave-like properties when not. They would be so light and numerous that it would collectively be equivalent to a luminous ocean of energy, radiation, and light. It will be described in physics as scalar-field dark matter (SFDM), similar to the Higgs field in the ordinary matter universe. Since the wave nature of these photons dominate, they are less point-like and are smeared out over spacetime. For this reason, they are also known as "fuzzy dark matter." According to Dr John White and Dr Stanley Krippner a 'Universal Energy Field' permeates all space, animate and inanimate objects; and follows the laws of harmonic inductance and sympathetic resonance. This scalar dark matter field is the universal energy field that metaphysicists have been talking about. Since there are four particulate spheres, there would be four universal energy fields, interpenetrating each other.

The energy contained in the fields of these photonic particulate spheres, makes up almost 70 per cent of the matter in the universe. While the physical-etheric, and the 3d-correlates of the 4d astral, 5d mental, and 6d spiritual plasmaspheres of the Earth can still be linked with the ordinary matter Earth, these higher dimensional photonic spheres exist in larger scale structures, within the largest universes in the local multiverse. The mystic Paramahansa Yogananda, says, "Souls in the causal world recognise one another as *individualized points* of joyous Spirit." "Individualized points" alludes to identity particles.

There are identity particles in every sphere, connected by virtue of the fact that they are different excitations of a single closed string (called the "Sutratma thread" in Hindu metaphysics.) According to M-theory, closed strings are not confined to any brane or universe – they are inter-dimensional objects. A string is extremely small, being only one Planck length – the smallest length possible in spacetime, i.e., about twenty-five orders of magnitude smaller than an atom. In higher energy and extremely low density particulate spheres, these identity particle-waves exist freely. However, in lower energy spheres, where particle density is high, they are obscured by the macroscopic plasma, photonic-plasma, or photonic bodies. When the macroscopic body dies, the identity particle exits the body and exists freely as a particle-wave in the higher dimensional particulate spheres where time is almost absent.

It subsequently becomes dormant when a new identity particle becomes active through the excitation of the tiny string at a new frequency, within the next macroscopic body that inhabits a different universe. When one excitation of the string (as defined in superstring theory) subsides and becomes dormant, and another excitation occurs within one unit of Planck time (an extremely short instant of time,) the locus of consciousness appears in the new particle caused by the new excitation – wherever the second identity particle is. This is similar, but more complex than the correlation of states that is observed in quantum teleportation, when a pair of quantum particles are separated while in an entangled state.

A more traditional analogy would be of one lighted candle lighting up another candle and then being extinguished. A modern analogy would be of a player taking off his VR (virtual reality) headset at the end of one virtual game and dons a new headset at the start of another. (You can bring the credits or debits of your score from one game to the next.) As noted earlier, your body (including the brain) is the VR headset, by analogy. From the coarse classical perspective, however, a smooth transition of consciousness occurs between macroscopic bodies. If no new excitation of the string occurs after the identity particle becomes dormant, there will be no longer any related stream of consciousness within spacetime. The score is settled, and the game is over.

Deep meditation simulates a near-death experience when it reduces all vital signs, for e.g., reduced heart and respiration rate, with sensory and cognitive suppression. In this state, the identity particle can be deactivated by advanced meditators. The identity particle of a higher energy body will then be spontaneously activated. They can subsequently deactivate it to transfer the locus of consciousness back to the lower energy body. The locus of consciousness can therefore migrate stepwise, from one higher energy body to the next.

Hindus and Buddhists might interpret the photonic particulate spheres, discussed in this chapter, as the "arupa" or "formless regions," as there are no macroscopic bodies or forms – only microscopic particle bodies with a probabilistic measure of existence. Buddhists generally found it difficult to understand how consciousness could exist without bodies, as it was taught that (discriminatory or dual) consciousness was interdependent with the body. Furthermore, since the conscious agent has not yet been liberated from conditioned existence, it would not be possible for the agent not to have a body of some kind (since the body is the karmic load of the agent.) There was some speculation in the Pali canon by the early Buddhists that the bodies must be extremely tiny.

The bodies are in fact extremely tiny, being elementary material particles that are invisible to the naked eye. When they are not observed, they exist as invisible waves, and which can only be described by the mathematics of quantum physics. (Nevertheless, they are material.) Pre-twentieth century Buddhists would have found it difficult to describe these as an understanding of quantum physics is required. The interactions of these conscious identity particles are quantum-like, with superpositions, non-local interactions, and quantum-tunneling between universes, occurring. Subjectively, advanced meditators and near-death-experiencers (NDEers), whose locus of consciousness is transferred here, will float in an empty space without observing any personal body.

7d-particulate sphere – pure space

As discussed in chapter 11, if we take the ascending energy pathway, we expect the time dimension to revert back to a spatial dimension incrementally. By the time we reach the 7d sphere, though, the effect of time will be negligible. The dominant experience will be that of "pure space." The speed of causality and the speed of light would be about half its speed in the 3d ordinary matter universe.

8d-particulate sphere – pure consciousness

According to Yogananda, these particulate spheres are indescribably subtle. In order to reach these spheres, you would have to possess tremendous powers of concentration; so that if you closed your eyes and visualized everything in the multiverse, you would realize that they exist only as ideas or information, and these themselves are projections of consciousness. Death and rebirth will be in thought only – like in a virtual reality game. In the 8d sphere, the dominant experience will therefore be that of "pure consciousness."

It is interesting that in the last few decades, there has been a growing consensus that information theory is of fundamental importance to physics. The legendary physicist, John Wheeler, coined the phrase "the IT from BIT." It implies that physics, particularly quantum physics, isn't about reality, but only the best *information* we have about it, based on the questions we ask. The metaphysical evidence suggests the same, but it goes further. It says: "IT from CHIT." (Chit means "pure consciousness" in Sanskrit.) Pure (non-manifesting) consciousness precedes discriminatory or dual consciousness. The latter appears when mathematical structures are generated in pure consciousness. It is the Platonic home of all mathematical structures and forms, which generates information, i.e., the physical laws that produced and governs the multiverse, and boots-up the virtual game of existence. All matter, energy and even spacetime are constructions or activities of this *universal/multiversal* discriminatory consciousness. *Individual* discriminatory consciousness, as mediated by the identity particle, further modifies this by introducing a mathematical structure which generates a "self" within spacetime. The structure weakens when the particle-wave leaves the macroscopic classical body and collapses when it becomes dormant.

The author believes that isolated elementary particles or collections of particles in the same quantum state, within the virtual reality of spacetime and matter, would manifest intermediate or mixed states of consciousness as an artefact of these constructions within consciousness. (These particles are not producing consciousness. Consciousness emanates from them.) As an analogy, when water partially freezes into ice structures, there would be ice crystals with water in it, and water with ice crystals in it. In a VR (virtual reality) analogy, the person who is wearing the headset will be experiencing pure consciousness, as well as its constructions. Hence, we would find artefacts of consciousness in the constructions. This idea has some similarities to panpsychism with regards to the observable outcome, i.e., particles with consciousness, although it starts from a fundamentally different place. To this extent, panpsychism can be reconciled to the idea of pure consciousness being fundamental and outside spacetime (as in philosophies such as Advaita Vedanta.) While panpsychism starts its analysis from matter, Advaita Vedanta starts from consciousness.

Summarizing, it can be seen that, in the ascending energy pathway, the multiverse can be seen as a continuum from matter to radiation (i.e., light or "LIT,") from information to discriminatory consciousness, and finally to pure (non-manifesting) consciousness. We could therefore say: "IT from LIT from BIT from CHIT."

9d-particulate sphere – pure nothingness

Each succeeding and expanding bubble universe (in the ascending energy pathway) will be older (as they were nucleated and created earlier,) and also inflate faster and earlier than lower-energy universes (see chapter 11.) This means they would have already petered out into almost nothingness. Therefore, as we ascend to the highest energy levels, the void will increasingly dominate in the universe. Additionally, as the speed of light slows down, the Hubble sphere (also known as the causal sphere or the sphere of causality in physics) contracts. Only a "void" exists where light from other parts of the universe cannot reach. In the 9d-particulate sphere, the conscious agent will therefore experience a void, or "pure nothingness." The dominant experience would therefore be one of "pure nothingness." This is not a subjective void. Anyone who is in this universe will experience it.

10d-particulate sphere – virtual individualized existence

The 10d sphere, which is largest in volume and highest in dimensionality, is at the "edge" of the manifested world, i.e., at the event horizon of the local multiverse. As the time dimension rotates to become a spatial dimension fully, and the speed of light falls to zero, the Hubble sphere contracts to a Planck volume. It will take an infinite amount of time for a signal to propagate from one point in space to another. Here, we will personally experience a "block universe," where there is only space and no time. However, it is not static. The whole history of the multiverse can be discerned instantaneously. The distinction between past and future would no longer be relevant, and causality (including what religions call karma) becomes inoperative.

This inoperability is only experienced by inhabitants in it. Causality will still operate in the lower energy spheres within it. Gravitons, being closed strings, are not constrained. Hence, shadow gravity from all these particulate spheres, which make up 70 per cent of the mass of the universe, will be felt by lower-energy universes. Since the identity particle-wave is *quantum* in nature, it would experience a *"superposed* block universe" where all futures and pasts allowed by the laws of quantum physics would co-exist.

High Energy Quantum Vacuum

Yogananda says beings live in the 10d formless sphere for thousands of years, and then by deeper ecstasies withdraw from the "little causal body" (interpreted here as the identity particle) and put on the vastness of an infinite-dimensional reality – the higher energy quantum vacuum. They no longer experience their joy as individualized particle-waves but are now integrated with the luminous cosmic "ocean." This is universal (not personal) discriminatory consciousness. As an ultra-high-energy state, it is a full-void, where dualities are no longer relevant.

From a particle-wave perspective, the identity particle devolves into a virtual particle, which pops in and out of existence, between the sphere and the high energy quantum vacuum. (Real particles can become virtual particles. For e.g., a real neutral pion may emit a photon, and turn into a virtual pion.) As a conscious virtual particle, it will alternatively perceive and not perceive the universe (now a void in a Hubble sphere with Planck volume.) After some time, without willing or thinking (i.e., through a natural process) all individual discriminating conscious activity will cease, and the particle will further devolve into a wave and subside into the high energy quantum vacuum of universal discriminatory consciousness. The equivalent description, from a string perspective, is that all excitations of the identity string ceases.

According to Taoism, the Tao is an endless void but also of universal energy, from which all things emanate. Conceptually, this is almost identical to science's high-energy quantum vacuum in eternal inflation theory. It is both an infinite-dimensional (high-energy) full-void where all dualities are held in check, as well as a zero-dimensional empty-void (vacuum) – the abode of pure consciousness where nothing manifests. The Hindu mystic, Yogananda, explains, "The soul realizes it is Spirit in a region of light-less light, dark-less dark, thought-less thought." The Buddha, Siddhartha Gautama notes, "Here long and short, coarse and fine, fair and foul, and name and form [i.e., dualities] are all brought to an end." He elaborates that there are no cycles of dualities in time, yet it is not frozen or static, "there, I say, there is neither coming, nor going, nor stasis; neither passing away nor arising." Space, time, and a personal self are no longer relevant, as in that reality, he explains, "[one] does not imagine he is aught or anywhere or anything." Christianity reveals: The Word (the full-void, the Logos, containing all mathematical structures) was with God (the empty-void, pure consciousness), and the Word is God (they are one and the same.)

Just like the ordinary matter universe, which is expected to accelerate its expansion and become less dense as more space is created, all universes will merge into the high energy quantum vacuum, starting with the highest energy universes, and moving downwards to lower energy universes. Hindus and Buddhists would interpret the whole of the multiverse of bubble universes, structured by spacetime, as "samsara." The natural trajectory of the conscious agent into or out of the high energy vacuum recapitulates the trajectory of the local multiverse which dissolves in time into the high energy vacuum. Information about each string of identity particles will be encoded on the cosmological event horizon of the multiverse, while the Tao reverts back to its original state.

Conclusion

The hierarchy of bubble universes can be conceived as different phases of a projection from a high energy quantum vacuum. Yogananda realized this in 1946 when he said, "A cinematic audience may look up and see that all screen images are appearing through the instrumentality of one imageless beam of light. The colorful universal drama is similarly issuing from the single white light of a Cosmic Source...One's values are profoundly changed when he is finally convinced that creation is only a vast motion picture; and that not in it, but beyond it, lies his own reality." The photonic particulate spheres map the region through which this cosmic beam of light travels. The light hits the first two "screens" in the 5d and 6d photonic-plasma universes, and thereafter dark plasma and ordinary matter appear on a further two screens in the 4d and 3d universes.

As we ascend, our bodies become more subtle, before disappearing from the projected multiverse, like a fading mirage over a black hole. In fact, there are many characteristics in the trajectory, during the exit from the multiverse and spacetime, which are similar to falling into a black hole. Time flips into space, the panoramic view of all the pasts and futures of the multiverse, followed by total darkness in a void, and then an inexorable flow towards an ineffable singularity (which Buddhists might identify as a "stream-enterer,") in a state called "Nirvana" by the Buddhists or "Moksha" by Hindus when it is reached.

The singularity may be characterised in many ways, including the Tao, Brahman or the unmanifest God. The final phase of the escape from the multiverse is therefore similar to the entry of a conscious agent into a black hole and exiting from spacetime altogether. Alternatively, it could be seen as the unmanifest God taking off his/her VR (virtual reality) headset when the game of spatio-temporal existence is over.

CHAPTER 18

❋

Dark Plasma UFOs

> ...be careful not to confuse manifested systems, existing on different planes..., with the far-flung solar systems and galaxies on the physical [i.e., ordinary matter] plane with which astronomers deal.
>
> I K Taimni, 1974

Taimni, a professor of chemistry at the Allahabad University in India and a metaphysicist, cautioned that astrophysical measurements made by scientists are all on the physical (i.e., ordinary matter) plane. It is therefore possible to establish communications with them through our ordinary matter sense organs and telescopes. The objects and entities from the physical-etheric, astral, and higher energy universes, however, are composed of different elementary particles and are therefore quite beyond the scope of our ordinary matter instruments (except for gravity.) Direct access can only be obtained through the subtler bodies and levels of mind that are linked to the ordinary matter body.

UFOs/UAPs Originate from Neighboring Dark Earths

Most UFOs/UAPs originate from the physical-etheric Dark Earth, which is closest in energy levels to the ordinary matter Earth, and more rarely from the astral Dark Earth. One of the most persuasive reasons as to why this would be more probable, than UFOs from other star systems, is the near-impossibility of ordinary matter vehicles to travel these vast distances within a reasonable time. Even in the slim chance that this would be possible, it would not be cost-effective for *intelligent* aliens to do so, as the benefits of visiting a relatively technologically primitive human civilization would hardly be worth the huge cost. It is easier for any dark plasma "alien" from a counterpart Earth in the physical-etheric universe to visit us than someone from another stellar system in our ordinary matter universe. (As Taimni cautioned us, above, we should not confuse the different types of universes.)

It also makes more sense when we consider the messages from UFOs (and also Marian apparitions) about imminent disasters and admonitions to take care of the ordinary matter Earth. Why would an alien from another stellar or galactic system be bothered about the destruction of this tiny rock? However, it would be understandable if these were from counterpart Dark Earths. They would be concerned about environmental degradation, nuclear holocausts, and the destruction of the ordinary matter Earth, as it an important training school for many plasma life-forms. More importantly, it may also impact their existence as we share the same gravitational field.

We will realize from the evidence that most of the encounters with aliens and UFOs are really sporadic encounters with inter-terrestrial beings from Dark Earths (not extra-terrestrials.)

> There is no indication that the UFO phenomenon is extraterrestrial. Rather than being from other star systems, there is mounting evidence that UFOs come from a multiverse which is all around us, and of which we stubbornly refused to consider in spite of the evidence available to us for centuries.
>
> Jacques Vallee, Dimensions: A Casebook of Alien Contact

Access to each of these Dark Earths gives us access to the infinite universes that they inhabit, and the diverse types of alien life-forms that could be encountered. Considering the huge volume of even one of these Dark Earths and the multiple biospheres in various shells, we can expect an enormous diversity of UFOs, including biological UFOs, and alien life-forms. The civilizations that are thriving in Earth's dark plasmaspheres and the societies in various heavens that are located in the atmosphere, longer than the human species in the ordinary matter Earth, are highly advanced in technology, including communications and transportation.

Production of UFO/UAP Phenomena in the Atmosphere

Kenneth Ring, Professor Emeritus of Psychology at the University of Connecticut, believes that these encounters are "imaginal experiences" — experiences which are midway between mind and the "hard" physical (i.e., ordinary matter) world. He says, "It is a real but mind-created world that individuals experience during near-death experiences...in subtler dimensions."

Most UFO/UAP sightings are located in the atmosphere and are crystallized thought-holograms within the nearby physical-etheric Dark Earth. They are generated by the interaction of trillions of thought-modulated dark electromagnetic waves broadcasted by millions of dark etheric bodies, belonging to contemporary humans who watch movies and TV, and the matter in this dark plasmasphere. These same processes produce religious heavens and hells, as discussed in chapters 14 and 15. Apart from the "natural" production of proto-intelligent plasma formations, such as dark plasma UFOs and UAPs, these waves can also mould and shape the appearance of intelligent beings from the physical-etheric Dark Earth. In this way, physical-etheric entities (with no corresponding ordinary matter bodies) and the dark plasma bodies of humans (who are asleep) can easily assume the shape and character of "aliens" — seen in the movies or TV. So, there are two levels of interaction: firstly, with proto-intelligent plasma holograms, and secondly, with intelligent plasma life-forms.

As discussed in chapter 14, just like the astral Dark Earth, the physical-etheric Dark Earth is filled with trillions of life-forms. Each shell in this dark plasmasphere represents a biosphere that is much larger and more extensive than the single-shelled biosphere on the surface of the ordinary matter Earth. The intelligent life-forms here would have their own governments, societies, cultures and advanced technologies. As discussed in chapter 10, communication mechanisms that can be used by these intelligent plasma aliens include the use of dark radio and acoustic waves in plasma, thought-holograms, and plasma bubble-drives. Transport mechanisms include electric propulsion using Birkeland currents, as well as mechanical propulsion using electric wind, plasma, and photonic thrusters. These aliens naturally use dark electric fields, plasma, or masses of photons, since these are abundant in their natural environment, just as squids use "thrusters" that issue water and birds use air currents to navigate.

The British Ministry of Defense (MOD) completed a four year study (1996 to 2000) which looked at data compiled from reports of UAPs received by the MOD over a 10 year period (1987 to 1997.) The study concluded that most of these sightings were due to ordinary *plasma* bodies generated by incompletely burnt-out meteors plunging into the atmosphere during meteor showers, although with certain reservations. They cautioned, "It is not certain that the radiation/fields are conventional and electromagnetic in nature...any pursuit of this process of identification or elimination is pointless if it turns out that UAP radiation is other than *EM [electromagnetic] radiation as we currently understand it*" (emphasis added.) Indeed, dark plasma life forms radiate dark (not ordinary) electromagnetic waves – as discussed in chapter 10. The MOD detection data provides evidence of dark plasma thought-holograms and more rarely, intelligent beings from Dark Earths, which become visible in Niche 3 of the neighboring physical-etheric Dark Earth, when they generate ordinary plasma.

UFO Abductions

Based on the evidence, most UFO related "abductions" most likely relate to our physical-etheric dark plasma bodies inhabiting the lowest energy physical-etheric Dark Earth, and not the ordinary matter body. There are usually no signs on the ordinary matter body and no ordinary matter objects left by aliens. Telepathy, the most common form of communication between subtle dark plasma bodies (as discussed in chapter 10,) is frequently associated with these abductions. Researchers have also noted that these "abduction scenarios" have many elements in common with "astral travelling" and "near-death experiences."

Detection of UFO/UAP Phenomena

Dark plasma aliens, who use dark plasma UFOs, can register on radar temporarily if ordinary plasma is generated for a short duration through the dark ionization process or when dark matter particles (such as axions and dark photons, and their equivalents) convert to ordinary matter particles (such as ordinary photons at radio wave bands) in certain conditions. If the ordinary plasma's frequency is lower than the ambient light in the atmosphere, it will absorb electromagnetic waves to become transparent and not register on radar. If it is equal to the frequency of the ambient light, the ordinary plasma will be sighted as a dark shadow. If it is higher, it will reflect electromagnetic waves, and it will look like a metallic object. (These appearances are the result of well-known plasma dynamics, discussed in chapter 9.) This ordinary plasma is what causes many of the UFO and UAP (Unidentified Aerial Plasma) sightings, which cloaks the dark plasma objects and beings.

When a person is falling asleep, the locus of consciousness shifts to the dark physical-etheric body. Conversely, when waking up, it shifts back to the ordinary matter body. At these times, sightings of dark plasma UFOs can be expected. People who practice regular mediation would also have access to the sensory system of the dark physical-etheric body and would therefore have a higher probability of sighting these phenomena. UFO hunters should take up meditation as this will increase their chances of experiencing more sightings to enable them to study UFOs/UAPs more closely.

CHAPTER 19

❈

Plasma Deities and Angels

Public Religious Apparitions

Public religious apparitions generally emanate from the lowest energy dark physical-etheric plasmasphere, although it may originate from higher energy dark plasmaspheres. There are human-linked plasma beings, currently residing in classical heavens, who have a genuine desire to interact with human beings on Earth to accelerate their "development." They would also be concerned and warn about any catastrophic events that could take place. (Note that the "present" in higher-dimensional plasmaspheres and spheres encompass our future – as discussed in chapter 11.) They manifest for short periods within our physical-etheric plasmasphere, by using the collective thought-holograms generated by human dark plasma bodies. It is theorized that apparitions of deities and angels in the ordinary matter Earth historically have originated from the classical heavens in Niche 4 in the astral plasmasphere – as they fit the profile of the inhabitants of that niche. (These classical heavens are discussed in chapter 15.) The lowest heavens in Niche 3 in the etheric plasmasphere is a staging area for most of these apparitions.

The manifestation of these astral deities and angels is done by first acquiring a physical-etheric body when they arrive in the physical-etheric heaven in Niche 3, and then increasing the density of this body by decreasing its volume and ingesting more physical-etheric dark matter particles. This makes it easier for humans, who already possess physical-etheric bodies, to see them. Ordinary plasma would be generated by these apparitions by various processes (including the dark ionization process and conversions of dark matter to ordinary matter particles that are allowed by physics, as discussed previously.) This could be detected using our ordinary matter bodies and scientific instruments.

This task appears to consume a lot of energy for these beings and presumably, based on anecdotal evidence; they would need to obtain relevant permission from the relevant authorities in these classical heavens, to contact humans. While we see them as aliens, they too see us as aliens. Religious apparitions are in fact visitations by aliens. Niche 3, in which religious apparitions occur, is the same region in which plasma holograms of UFOs and UAPs manifest. They are in fact the same type of phenomena – only the characters differ. The narratives of both are sometimes, in substance, quite similar. Usually, this includes a warning about some imminent catastrophic danger and for humans to mitigate the risks. Hence, permissions are tightly controlled and appearances of high-ranking astral entities here are rare. If they are residing in a pleasurable classical heaven, they may not be motivated to abandon that. It therefore constitutes a considerable sacrifice on their part to "come down" (literally, in terms of energy levels) to communicate with us on something that they think is important and urgent. However, they are not omniscient. They are evolving in Earth's plasmaspheres, as much as we are evolving in this sphere, due to natural universal forces.

In the twentieth century, there were 386 reported cases of Marian apparitions. Popular writer, Michael Talbot, says that Marian visions are not appearances of the historical Mary, but psychic holographic projections created by the collective unconscious. Where more intelligent behaviour is identified, however, it is possible that these are real living entities from higher energy plasmaspheres whose bodies are shaped (rather than created) by the thought-holograms of the collective unconscious. So, there are two levels of interaction: firstly, with proto-intelligent religious plasma holograms, and secondly, with intelligent plasma life-forms from dark sectors.

It is therefore possible that the Marian and related apparitions relate to the historical personality of the first century Jewish Mary, the mother of Jesus. During the Marian apparitions, Mary arguably played the role of a Bodhisattva, in a Catholic setting – and is sometimes characterized as such by Buddhists, in the form of the goddess of mercy, "Guanyin." The classical Catholic heavens, in Niches 3 and 4, and the corresponding classical Catholic hells, in Niche 1, among others, form a single structure within the Catholic BST (Belief System Territories.) Catholics will transmigrate to these locations after the death of their ordinary matter bodies.

Plasma Deities and Angels

Plasma angels and deities would have the general properties of plasma bodies, as discussed in chapters 9 and 10. For e.g., they will emit a gentle light in the glow mode of the plasma and a bright intense light in the arc mode. They would be thermochromic and could therefore be very colorful. They will have features associated with the Sun — besides displaying coronas, a closer inspection may reveal granulations, spicules, and striations on their bodies. They radiate dark electromagnetic and acoustic waves, allowing them to communicate their mental and emotional states, and messages, to the dark plasma bodies of human observers – without actually speaking. They can be seen as bright balls or orbs in their natural state (as in the Marian apparitions, discussed below.) Occasionally, they can take on a humanoid appearance temporarily, using a process with similarities to cymatics and plasma holography technologies, to communicate with humans. This has been discussed in chapter 10.

Case Study One — The Zeitoun Apparitions

The report of the General Information and Complaints department, Zeitoun, Egypt, in 1968, states that "Official investigations have been carried out with the result that it has been considered an undeniable fact that the Blessed Virgin Mary has been appearing on Zeitoun Church in *a clear and bright luminous body* seen by all present in front of the church, whether Christian or Moslem" (emphasis added.) This apparition was caught on camera and official TV. Images show a translucent body. As noted in chapter 9, plasma bodies can become transparent or translucent, depending on the internal plasma frequency, relative to the frequency of the external ambient light. They generate light through spontaneous emission. Dark plasma bodies can generate ordinary plasma and photons which would allow cameras to capture the images, and many people to see the apparition using ordinary sight.

Case Study Two — The Fatima Apparitions

Moving Plasma Balls and Transparent Bodies

Fatima is now well known to the Catholic world. "Our Lady" first appeared to three small children tending sheep near Fatima, Portugal on May 13, 1917, and asked that they return to the site on the thirteenth of each month until October. Before this, in the spring of 1916, the children were tending their sheep at a rocky knoll not far from their home. A *sudden strong wind* on a calm day startled the three peasant children out of a game they were playing, and they saw across the valley a *dazzling globe of light like a miniature Sun*, gliding slowly towards them. As it approached, the ball of light gradually resolved itself into a *brilliantly shining* young man. According to the children they saw "a light whiter than snow in the shape of a *transparent* young man, who was more brilliant than a crystal struck by the rays of the Sun" (emphasis added.) He identified himself as the "Angel of Peace" and enjoined them to recite a prayer. Then he disappeared by fading away. On September 13, 1917, an assembled crowd saw a *bright ball* in the sky coming from the east, which glided majestically into view for a few seconds but then vanished, as far as the crowd was concerned. The children, however, see a *luminous globe* hovering on a little oak tree. The Virgin speaks to the children then returns the way she had come — in the shape of an easily visible *luminous ovoid* moving away in the direction of the Sun, "calmly but with a certain speed," before disappearing.

The above excerpt from the sources contains numerous descriptions that suggest plasma activity. A *strong electric wind* can be generated by a plasma body, as well as the sound of thunder. (See chapter 10 for details.) The default shape of dark plasma bodies is an ovoid, i.e., like a *ball*. In the arc mode of plasma, it shines with a *dazzling light*. These bodies can change their degree of opacity to become *transparent*. (See chapter 9 for details.) Jacques Vallee has argued that the moving "glowing globes of light" can be interpreted as vehicles and were regarded as such by some of the people who saw them. As noted in chapter 10, plasma bodies (which are usually enclosed in plasma ovoids) use electric propulsion, powered by Birkeland currents and electric winds, as well as use plasma and photonic thrusters. Several researchers have pointed out that the Marian apparitions have many features of UFO encounters.

Sometimes, only the children (who were in a state of prayer for a long time) could see the apparition. In this case, the locus of consciousness had been transferred to their physical-etheric dark plasma bodies, allowing them to use dark sight to see the physical-etheric body of the being. Other people in the crowd, who have done the same (i.e., those who have been in prolonger prayer or have otherwise activated their physical-etheric body,) would also be able to see the dark plasma being, while the rest, using only their ordinary matter bodies, would not. The perceptions from different sensory systems would generate different reports of the event.

Miracle of the "Sun" in Fatima

On October 13, 1917, when Mary was about to leave, she pointed to the Sun. The rain stopped, and the clouds parted, and the "Sun" then began to whirl in the sky, scattering rays of multicolored light, lighting up the entire countryside. It whirled for three minutes, stopped, and then whirled again a second and third time lasting a total of twelve minutes. Then it spun faster each time and at the end seemed to tear itself from the sky and began plunging to Earth. Many in the crowd thought that it was the end of the world. Other people witnessed the solar miracle from a distance, ruling out the possibility of any type of collective hallucination. The heat of this "Sun," as it descended on the people, was reported to have had the effect of drying their clothes and the ground, so that they went from being completely soaked to being dry in about ten minutes.

This "Sun" appeared to be an objective phenomenon, as many people saw it, even outside the immediate location — but was it the Sun? If the Sun descended as quickly as reported it would immediately vaporize the Earth — it would not take ten minutes to simply dry clothes or the ground. Within eight minutes, Earth would be spun out of its orbit — and so would other planets at different times. There have been similar observations, which had nothing to do with religious apparitions. In 1923 Joseph Mintern saw the "Sun" behaving in an unusual fashion, surrounded by flashing rays and changing colors; and appeared to dance and shift about from here and there. "The colors were so brilliant and dazzling that even after I had come indoors anything I looked at appeared in a mixture of all the colors seen," Mintern said. Mintern described this "Sun" as a "kaleidoscopic Sun." In June 1885, several residents of Birmingham, Michigan, saw a similar display just before sunset. Interestingly, they also observed "innumerable balls of decomposed light the size of bushel baskets" in the sky and on trees.

According to physicist William Corlis, the laws of optics predict that atmospheric ice crystals that cause common halos will also produce bright patches of light to the right and left of the true Sun on the halo ring. "Such mock suns and mock moons are not at all unusual when halo conditions prevail," he says. However, he contends that the "kaleidoscopic Suns" as observed by Mintern, and others, are "well beyond the capacity of atmospheric optical theory to explain." He also goes on to say, "The similarities between the 'dancing Sun' of Fatima and kaleidoscopic Suns are strong."

The vision at Fatima was not of the Sun, but a high energy physical-etheric plasma ball (similar to a UAP) nearer to the Earth, which generated ordinary plasma (through processes already discussed.) It would be in the troposphere (the lowest part of the atmosphere) — generated by the weather in the physical-etheric Earth or engineered by the dark plasma beings inhabiting it or higher energy beings passing through it. It would be very much like the Sun because the Sun is in fact a ball of plasma; and it would generate heat. At the Fatima apparition of May 13, 1917, a "young man" appeared out of a dazzling globe of light which was described by the seers as "a miniature Sun." These plasma ovoids are native to dark plasmaspheres.

Ordinary plasma balls are known to exist in the ordinary matter Earth's ionosphere. The Sun photo-ionizes the upper atmosphere when the Earth is facing it. When the Earth rotates away from the Sun, some regions of the plasma in the ionosphere cool down. These cooler regions then separate from the rest of the plasma because of the difference in temperature and float in the ionosphere as plasma balls. (As noted earlier, plasmas with different properties naturally separate and generate plasma sheaths around themselves.)

Vision of Hell at Fatima

On July 13, 1917, in Fatima it was reported that Mary opened her hands and rays of light from them seemed to penetrate the Earth allowing the children to see a terrifying vision of hell, full of demons and lost souls amidst indescribable horrors. This vision was so real that the children later made severe sacrifices for the salvation of sinners. Mary responded to the vision with sadness and tenderness. Lucia Santos, the eldest of three children, reported in 1941 the horrifying scene as follows, "Plunged in this fire were demons and souls in human form, like *transparent* burning embers, all blackened or burnished bronze, floating about in the conflagration, now raised into the air by the *flames* that issued from within themselves together with great clouds of smoke, now falling back on every side like sparks in a huge fire, without weight or equilibrium, and amid shrieks and groans of pain and despair, which horrified us and made us tremble with fear. The demons could be distinguished by their terrifying and repulsive likeness to frightful and unknown animals, all *black* and *transparent*" (emphasis added.)

As noted previously (see chapter 9,) plasma bodies are able to change their internal frequencies to appear black or transparent. Furthermore, as discussed in chapter 14, hell is indeed a loathsome and unpleasant place. However, it is temporary and more for the detoxification of the current dark plasma body. Since the plasma body itself is relatively hot and could actually be hotter than most carbon-based fires, inserting a dark plasma body into "fire" will not make much difference. The use of the terms "smoke" and "flames" in the description indicates a medieval-based carbon-based fire. A more modern society might have used microwaves or electronic incineration methods. So, these hells seem more like period dramas cast in dreamscapes. The generation of these dreamscapes are discussed in chapters 10 and 14. Furthermore, it is interesting that the vision was generated from the rays of light that issued out of Mary's hands. This suggests that it could be a holographic projection, using plasma holography and photonics.

Other Plasma Formations in the Sky

Plasma physicist Anthony Peratt has demonstrated that a giant plasma column was produced in the atmosphere of the Earth some four to five thousand years ago and that it was luminous enough to be observed by human populations around the world. Some plasma physicists are now comparing electrical plasma discharge formations in the laboratory to rock art images around the world which suggest that immense and terrifying plasma configurations were seen in the sky thousands of years ago. These are analogous to today's UAPs.

Ancient myth-makers would not have the knowledge to recognize a plasma discharge at a sufficient distance from the Earth. For e.g., the Chinese would most probably interpret a plasma discharge as a mythic dragon that "breathes fire" and displays "long-flowing hair" or "feathers" — all tell-tale features of a plasma discharge. The "writhing form" that appears in the sky and the darkness which follows are what we should expect if it was a discharge.

However, while it will be easy to assume that these ancient artwork and myths relate to plasma originating entirely from the visible, ordinary matter Earth, it once again suggests that, in fact, they were triggered by events in neighboring Dark Earths. As discussed earlier in this chapter, dark plasma formations may have occurred in Niche 3 of the physical-etheric Dark Earth, which can be viewed by humans in reverie, or in particular states of consciousness – both while going to sleep (hypnagogic state) and waking up (hypnopompic state,) and also in meditative states. These would generate transient ordinary plasma and photons, through processes already discussed, which can be viewed by a wider audience.

Other Plasma Phenomena in Religion

There are numerous dark plasma phenomena that can be easily identified in religious scriptures and apparitions. Here are a few examples. Firstly, when Moses, the most important prophet in Judaism, spoke and listened to a "burning bush" that did not seem to be actually burning in the Bible's Old Testament – he was communicating with a dark plasma being. When metaphysicist Leadbeater tried to describe plasma in 1910 (the term "plasma" was only invented in 1929,) he explained, "Try to think of a fire which does not burn." Today, we sometimes use the popular term "cold fire" for non-thermal plasma which has medical and industrial applications. It can be touched without any harm. Moses's "burning bush" can be explained by non-thermal plasma.

Secondly, when the Holy Spirit manifested at Pentecost, as recorded in the Acts of the Apostles in the Christian New Testament, there was a strong wind (i.e., an electric wind caused by a plasma body (see chapter 10 on how this is generated)). After this there was a fire overhead, which then distributed itself as "tongues of fire" above the heads of the apostles. Tongues or tendrils of fire would dart out when the main dark plasma formation is attracted electromagnetically to the apostles' heads, which would be acting as electrical conductors. These are basically plasma discharges. What appeared like doves are also seen in the baptism of Jesus on the banks of the Jordan river, as well as at the Pentecost, and more recently in many Marian apparitions. These can be easily identified as decomposed plasma formations from plasma discharges.

Thirdly, it has been recorded that Jesus appeared, wearing a white garment, with *red and pale rays* emanating from his heart, to the Polish Roman Catholic nun, Sister Faustina Kowalska, in 1931. Plasma beings can issue collimated beams of plasma from vortexes in their bodies which may have healing properties. These are similar to laser-induced cold plasmas that are used in medical applications today. They are also similar, in terms of mechanics, to a much scaled-down version of astrophysical phenomena, such as cosmic jets issuing out of rotating neutron stars.

Generally, angels have appeared in the Marian apparitions (discussed above) and in many other religious phenomena as bright balls of light – the most common appearance of intelligent dark plasma beings, and there are many other phenomena suggesting plasma activity. For e.g., St Paul was blinded by a bright light on his way to Damascus and Jesus became radiant during the transfiguration at Mt Tabor.

CHAPTER 20

❋

Dark Plasma Ghosts

Most of the ghosts encountered by us originate in the shells of the 3d physical-etheric Dark Earth, and less so from the 4d astral Dark Earth, which are coincident with the Earth's surface.

Dark Physical-Etheric Ghosts

The higher energy physical-etheric bodies may take a few days or several years after the death of the ordinary matter body to die. As long as it is not detached and/or dead, the person will not be aware of the next higher energy universe (usually the dark astral universe.) It will be "stuck" inside the physical-etheric Dark Earth — visiting familiar places. These etheric ghosts of human beings in Niche 2B of Earth's 3d plasmasphere are more frequently seen than astral ghosts in Earth's 4d plasmasphere because they are closer in energy levels to our world and inhabit the same dimensional structure. They include those whose ordinary matter body has recently died, as well as those who have already been detoxified in hell, after the death of the ordinary matter body. The etheric ghost may intentionally or unintentionally meet living persons who are awake in their (higher) etheric doubles, when the ordinary matter body sleeps. When they awake, most living persons will describe "dreams" where they met with their deceased. Some of the so-called "ghosts" may also include persons who have still not left their ordinary matter bodies completely (whom we normally refer to as the "living.") They would have higher density physical-etheric bodies because of the compactifying effect of the linked ordinary matter body (see chapter 9.) It may actually be quite difficult to distinguish them from their ordinary matter counterparts.

Dark Astral Ghosts

Although less common than etheric ghosts, astral beings residing in Niche 2C may also be encountered as ghosts on rare occasions. (This niche is coincident with the ordinary matter Earth's crust. See Table 2 in chapter 13.) It is rare because most astral beings prefer to live in more pleasurable heavens higher up. The astral plasmasphere is extremely huge and there are very few reasons why busy astral beings should come down to the crust or stay there. The probability would also be very low since the crust is a very thin slice (i.e., less than one per cent) of the Earth. The likelihood would be about the same as landing a dart straight into the bull's eye. The majority of people are also astral-blind. More people have developed lower frequency physical-etheric sight (consciously or unconsciously) than astral sight and would therefore more likely see physical-etheric ghosts. Life in Niche 3 is like life in the ordinary matter Earth, minus the ordinary matter body and its necessities, for astral ghosts.

Beings from Niche 4 (the classical heavens,) who have had accumulated heavy, low energy, dense dark matter in their bodies due to their lifestyles, while enjoying the heavenly realm, do fall into Niche 2C. These may include the arrogant and wrathful Buddhist Asuras and the like, such as the Christian Satan, who seems to have a similar personality profile as an Asura, and like them, have fallen from a heaven.

Properties and Characteristics of Plasma Ghosts

General Properties of Plasma Ghosts

The general properties of plasma bodies have been discussed in chapters 9 and 10. This is a recap, focused on plasma ghosts. Just like most plasma bodies, ghosts can move through each other, and walls. However, electromagnetic effects can be felt as their bodies glide near or through your own electromagnetic plasma bodies. Ron Cowen says, "evidence indicates that when speeding fragments of dark matter meet, they don't collide as other matter do but pass right through each other, ghostlike."

> This "dark matter" differs from ordinary matter in being able to pass right through both ordinary matter and other dark matter, just like ghosts are supposed to pass through stone walls.
>
> Theoretical Astrophysics Group,
> University of Oxford

Other general properties include the ability of ghosts to change their degree of opacity - they can appear and disappear by changing their internal plasma frequencies. Ghosts would also be able to communicate telepathically to some human observers through their dark plasma bodies. The transient ordinary plasma generated by them could generate electric, magnetic, and electromagnetic fields that can be picked up by instruments. The contact with them can be at different temperatures, as there is both hot and cold (or non-thermal) ordinary plasma. Ghost can also generate electric winds with the production of ordinary plasma.

According t0 Leadbeater, one difficulty of the recently disembodied is that they usually require considerable experience before they can clearly identify objects and navigate around them — just like a baby that needs to learn to walk. The stone-throwing, trampling, or vague movements of purported ghosts in haunted locations are attempts by recently disembodied persons to come to terms with their new environments. An unpleasant person in this life would probably be an unpleasant ghost, for some time immediately after his/her death, and a pleasant person, a pleasant ghost.

Plasma Ghosts and Electricity

According to the paranormal literature, ghosts appear to feed on electricity. Many ghost investigators believe that ghosts are sighted often in rooms or places near power lines or feeder cables coming into a building. Anything that produces a strong electromagnetic field seems to bring ghosts out of the invisible woodwork and into our living or bedrooms. This is also the reason why, perhaps, many ghost stories take place during electrical storms, says Craig McManus, a psychic medium and ghost hunter. "The electricity in the air seems to charge-up the bodies of ghosts in the area. Once charged-up, they can manifest themselves — usually for a short time, before the air becomes relatively neutral again."

It is more probable that an already slightly ionised environment, caused by electrical equipment, overhanging power lines, or natural events, helps ghosts to generate visible ordinary plasma more easily through the dark ionization process. This process can be triggered by increasing the density of the physical-etheric body which would increase the probability of collisions between ordinary ionized particles near electrical installations and dark matter particles.

McManus also notes that ghosts seem to have no problem manipulating electrical appliances. The computer goes crazy, the television flickers, the phone rings but there is no caller, and doorbells chime for no reason. A person who has recently lost his ordinary matter body but whose etheric double is still intact may be able to cause electrical and electromagnetic interferences by generating ordinary plasma through the dark ionization process to manipulate or short electrical circuits. Ordinary photons, generated through the natural conversion from axions and dark photons, can also affect electrical activity. The ghost may lack sustained mechanical control over objects but may have sporadic control over electrical and electronic devices — for example light switches and computers. Lights may go on and off without any mechanical change in the position of the light switch.

Similarly, messages may appear on the computer screen without any movement of keys in the computer keyboard. Some even claim that broadcasts by various beings from parallel spheres (most likely the physical-etheric sphere) have been transmitted and received by television or radio. This is conceivable — considering the ability of plasma bodies to broadcast dark radio waves, as discussed in chapter 10. Axions and dark photons, which are dark matter particles, can convert into ordinary photons. Dark radio waves can therefore convert to ordinary radio waves.

Detecting Ghosts

Dave Oester reports that when ghost researcher, Troy Taylor, took the first photo showing a glowing ball of light in the Bell Witch Cave of Tennessee, the ball was not visible at the time he took the photograph, but it appeared on film later. Similarly, scientists studying galaxies which are invisible to the naked eye, leave their cameras directed at them for long hours or days. The galaxies then "manifest" on a photograph, through the accumulation of photons. Hence, it is possible that objects, which are at the edge of our perception, can be captured on film if the camera is directed accurately at the target and given sufficient time to accumulate enough ordinary photons to show up in an image. The photons can be generated by the ordinary plasma produced by ghosts, during the dark ionization process. Dark matter particles, such as axions and dark photons, can convert to ordinary photons – so these can also be captured by camera, given enough time.

CHAPTER 21

❈

Beyond Earth

Dark Galaxies

Chung-Pei Ma, an associate professor of astronomy at UC Berkeley, and Edmund Bertschinger of the Massachusetts Institute of Technology (MIT,) say that our galaxy, the Milky Way, has about a dozen small ordinary matter satellite galaxies, but in computer simulations thousands of satellite galaxies of dark matter are seen. Furthermore, numerous dark galaxies have been detected using gravitational lensing. If ordinary matter follows dark matter, there should be nearly equivalent numbers of each, i.e., many more ordinary matter satellite galaxies. However, this isn't so, according to Ma. Some astrophysicists have suggested that this is because there are many more dark galaxies with very little ordinary matter than those that have.

Dark galaxies may have lost their visible ordinary matter counterparts due to collisions, gravitational attractions by larger bodies and other astrophysical events or simply because they did not fulfil all the conditions to attract ordinary matter in the first place, for e.g., there are numerous low mass dark matter halos that are too small to gravitationally attract ordinary matter. We would therefore expect many more dark galaxies (i.e., galaxies with very little ordinary matter,) than there are visible ordinary matter galaxies, with or without dark matter counterparts. This is because there is five to six times more dark matter than ordinary matter in the universe.

The mystic, Paramahansa Yogananda, noted in 1946: "just as many physical [i.e., ordinary matter] suns and stars roam in space, so there are also countless [dark] astral suns and moons." Based on the metaphysical evidence, we would expect numerous dark physical-etheric, astral, and higher energy dark stars, containing very little ordinary matter.

Dark Matter Planets within the Solar System

Planets in the Solar System follow Newton's law of gravitation closely. However, this does not mean there could not be a volume of dark matter in the Solar System that could support very low density radiation-like dark plasma life-forms. Dark matter counterparts of visible planets and dark planets will not impact gravity measurements in the Solar System significantly, as discussed in chapter 13. Physicists Foot and Silagadze, have written a paper arguing that there could be mirror planets within the Solar System. According to the metaphysical literature, the visible planets in the Solar System do have dark matter counterparts. Additionally, there are also invisible very low mass dark astral and higher energy dark planets, without any or insubstantial ordinary matter counterparts, within the Solar System. Metaphysicist Pearson reported in 1957, long before Vera Rubin and Kent Ford in the 70s ignited interest in dark matter in the scientific community, the following:

> So far, we have only considered the physical [i.e., ordinary matter] planets; but there are non-physical [i.e., dark matter] planets too ... [making up] by far the greater part of the planetary population of the [Solar] System.
>
> Norman Pearson, Metaphysicist

Metaphysicist Leadbeater had said in the early 1900s that all the astral counterparts of our Earth and of other physical (i.e., ordinary matter) planets, together with the purely dark astral planets, make up collectively the astral body of the Solar System. Similar conclusions can be made of higher energy dark bodies within the Solar System. In other words, there are physical-etheric, astral, and other higher-energy Solar Systems, with their corresponding Suns and planets, all interpenetrating each other. The physical-etheric, astral, and other higher energy universes have their own galaxies, stars, and planets — only a small number have ordinary matter counterparts that are visible to us in our universe. The estimated two trillion visible galaxies in the measurable universe, though staggering, represents only 1 per cent of the mass of the local multiverse! (Besides exotic dark matter; this percentage estimate also considers ordinary dark matter, for e.g., black holes, brown dwarfs, small planets, and invisible hot gas, as they are also generally not visible.)

Access to Other Dark Planetary Counterparts and Planets

The habitats of our neighboring planets in their dark matter extensions may be conducive to the related type of plasma life-forms, even if they are not conducive to ordinary matter biomolecular life. However, meeting them will be rare. According to Leadbeater, matter of the lower planes is never carried over from planet to planet. For e.g., the physical-etheric and the astral plasmaspheres of the Earth do not reach any ordinary matter planets. Hence, most human beings after death do not find themselves there. A person cannot pass from planet to planet in his physical-etheric or astral body than he can only using his ordinary matter body. Overall, therefore, contact with physical-etheric and astral beings in neighboring planets, such as Mars and Venus, would be rare.

However, there are certain special situations. Firstly, in a developed higher energy body from a higher plane (say, the 5d plasmasphere,) this may be possible by transiting through the Sun's much more extensive dark plasmaspheres which interpenetrates Earth's plasmaspheres (see next section for more details.) Secondly, dark plasma and photonic technologies may be used. After all, we have sent missions to the Moon with our ordinary matter vehicles which have traversed large expanses of empty space. Paramahansa Yogananda, in 1946, reported the use of astral vehicles and masses of light (photonics) to travel from one planet to another. Technology is therefore as relevant to societies in the physical-etheric, astral, and higher energy plasmaspheres of the Earth as it is relevant to us.

The Sun's 5d Dark Plasmasphere

More than 99 per cent of the (ordinary matter) mass of the Solar System is concentrated in the Sun. Its magnetosphere (also known as the heliosphere,) with its magnetic field, extends well beyond Pluto. The field overlaps with the Earth's magnetosphere and connects with its magnetic field. As noted above, the Sun itself has its dark matter counterparts (and therefore its own dark plasmaspheres,) i.e., Dark Suns. This has been observed and recorded by experimental metaphysicist, Helena Blavatsky, in the nineteenth century. She noted an "invisible Sun," i.e., a Dark Sun, behind the visible Sun.

The Sun's diameter is 1.4 million km (870,000 mi) or about 110 Earth diameters. The minimum distance of the Earth from the Sun is 146 million km (91 million mi,) and the maximum is 152 million km (94.5 million mi.) If the Sun's astral plasmasphere extends by about thirty Sun diameters (i.e., by the same proportion of the Earth's astral plasmasphere to the ordinarily visible Earth, which is about thirty Earth diameters,) then the 3d-correlate of the Sun's 4d astral plasmasphere should terminate at about a distance of 42 million km (26 million mi.) This is within the orbit of Mercury. The Sun's 5d plasmasphere would be 5.5 times larger (this is an arbitrary first approximation, based on the ratio of dark to ordinary matter in the measurable universe,) i.e., about 231 million km (144 million mi.) The Earth's orbit would then be within the 5d plasmasphere of the Sun throughout the year.

This opens up interesting possibilities. Although the Earth's astral plasmasphere is not extensive enough to reach other planets, an ambitious astral traveler who passes into it may be able to navigate within the Sun's 5d plasmasphere (which interpenetrates it) using his/her linked 5d dark plasma body, and then step down to a new solar 4d astral body to arrive within the Sun's astral plasmasphere. (The planetary 4d astral body cannot traverse the region, hence a new astral body is required.) However, most human beings would be confined to Earth's astral plasmasphere.

> Hiranyaloka inhabitants have already passed through the ordinary astral [plasma]spheres [on Earth,] where nearly all beings from Earth must go at death. None but advanced devotees are drawn by cosmic law to be reborn in new astral bodies on Hiranyaloka, the astral Sun or heaven, where I am present to help them.
>
> Paramahansa Yogananda, quoting
> his guru, Sri Yukteswar

This method can also be used by advanced travelers to reach the dark plasmaspheres of other planets within the Solar System. Unlike the difference in composition between the ordinary matter Earth (composed of atomic and molecular matter) and the ordinary Sun (composed of matter in the state of a plasma,) the difference in composition between the Dark Suns and the Dark Earths is much less – since both are made of matter in the state of a dark plasma. Travel within these plasmasphere with a dark plasma body or vehicle would be less problematic than travelling through ordinary matter space (composed of plasma) with an ordinary matter biomolecular body. In the latter case, they are composed of different states of matter and substances.

The Galaxy's Dark Plasmaspheres

Our galaxy, the Milky Way, is a significant member of the "Local Group" of galaxies. This group comprises over thirty galaxies, with its gravitational center located somewhere between the Milky Way and the Andromeda Galaxy and belongs to the Virgo Supercluster. This Supercluster has about one hundred groups and clusters of galaxies and is dominated by the Virgo cluster near its center. Our Local Group is located near the edge and is drawn towards the Virgo cluster. A study team at Johns Hopkins

University, using the Hubble Space Telescope and computer simulations, to study concentrations of dark matter in clusters confirm that they are located at the densest regions of the dark matter haloes. This shows that, beyond galaxies, there are also concentrations of dark matter in the centers of galaxy clusters. Hence, we would expect even the Virgo Supercluster, the Local Group, and our galaxy to have the 3d-double physical-etheric, the 4d astral and higher energy dark plasmaspheres – all embedded in the cosmic dark matter web. There will be other plasma, photonic-plasma, and photonic life-forms in these regions. However, contact with them, although theoretically possible, would be generally rare because of the great distances.

Heavens and Hells in Other Planets

Many, including scientists, are keen to know whether there are other Earth-like planets in the universe which support intelligent life. Astrobiology is the scientific study of the possible origin, distribution, evolution, and future of life in the universe, including that on Earth, using a combination of methods from biology, chemistry, and astronomy. *Dark astrobiology* (a new term used in this book by the author) is a branch of astrobiology that studies life-forms composed of exotic dark matter particles. It uses a combination of methods mainly from biology, dark matter, and plasma physics.

Data from the Kepler space telescope estimates that about 300 million potentially habitable planets could be in our galaxy. They could even be quite close (based on astronomical scales,) with several within thirty light-years of our Sun. Dark matter is pervasive throughout the universe, and so is plasma. We would therefore expect most life-forms, including intelligent beings, to be dark plasma and photonic life-forms. These millions of planets would have their own dark matter counterparts (i.e., dark plasmaspheres.) This means they will have their own ecosystems of heavens and hells, together with their own ecological niches, where different types of plasma, photonic-plasma and photonic lifeforms thrive and evolve. Just like our Sun, stars, too, would have their own 3d (physical-etheric,) 4d (astral) and higher energy plasmaspheres where plasma and photonic-based intelligent life would thrive.

However, we do not have to look far into other star systems for intelligent life. The dark matter counterparts of the Solar System, with its many extremely very low density radiation-like dark planets and dark matter counterparts of the visible planets, is rife with plasma-based life-forms. Mars, and Venus, our nearest planets, may have dark plasmaspheres harboring a variety of plasma and photonic-based life-forms. The most obvious and cheapest location to look for intelligent life, is actually in our own dark backyards (i.e., the Dark Earths,) the most accessible one being the lowest energy physical-etheric Dark Earth.

Currently, the human species is regarded as the only intelligent life-form on Earth. Some scientific researchers, searching for extra-terrestrial life, believe that it is the only intelligent life in the universe. This may need to be revised as there are numerous species of intelligent plasma, photonic-plasma and photonic life-forms occupying the various Dark Earths. Furthermore, the intelligence of many species of these life-forms far outstrips humans. A SITI (Search for Inter-Terrestrial Intelligence) programme may actually be more useful and cost-effective than a project on SETI (Search for Extra-Terrestrial Intelligence.) Many dark matter experiments may have unintended benefits in that they are actually feeble and unintentional attempts at SITI.

If there are more than 300 million ordinary matter planets with their own dark matter spheres in our galaxy, it is mind-boggling to think of the measurable universe as a whole, which has more than two trillion galaxies. Even this number only represents ordinary matter galaxies - dark galaxies are expected to outnumber them. We would still have to account for invisible dark matter planets without ordinary matter. Each of these trillions upon trillions "worlds" would have their own biospheres containing heavens, hells, and Earth-like niches.

Beyond our local multiverse, there are other multiverses with different histories. However, we would not be able to communicate with them, as there are no causal links. Many scientists now think that the measurable universe is only a small region of a larger, possibly infinite, universe. For all practical purposes, we are then forced to conclude that there are an infinite number of heavens and hells. The modern conception of an infinite number of "worlds" is beyond any religious cosmology and our capacity to even imagine. In this book, we have only briefly covered a rough sketch of the local multiverse – focusing only on the Earth-based environments where most human-linked activities occur. The time-scale is also restricted to the past ten thousand years, when the civilization relating to homo sapiens began to flourish.

Conclusion

Contacting Heaven

Science fiction films and novels have suggested that, through advanced technology, heaven can be reached by the living through conventional means. Such was the case in the Disney film "The Black Hole," in which the crew of the spacecraft found both Heaven and Hell located at the bottom of a black hole. Today, in the modern age of science and space flight, it is widely assumed that heaven is not a physical place in this universe. Religious views, however, still hold heaven as having a dual status as a concept of mind, but also possibly still a physical place existing on another "plane of existence," or perhaps at a future time. This book shows that heaven is *both* a state of mind with a corresponding body, and an actual location, in most cases linked to the ordinary matter Earth.

Transforming Hell into a Rehabilitation Centre

More religions are changing their concepts of hells into purification or detoxification centres, based on restorative justice. Catholicism introduced the concept of purgatory in the eleventh century. This is meant to be a purification centre, as a prelude to heaven. Evangelical churches are also now rejecting brutal hells and blind retributive justice. Some versions of Islam are now conceiving hells as purification centres. All religions should embrace the evolution of society in the twenty-first century, and the development of better human rights, to update their teachings.

The traditional hells, built by the trillions of thought-holograms of current and previous societies, should be abhorrent to anyone living in the twenty-first century. Instead of blaming God for these hells, or someone else, you can make a change. A simple rule: just forgive people who hurt or insult you. In that way, you will not be sending vengeful thoughts to the imaginal realm. Also, embrace science, to burn off the cob-webs of religious superstitions depicting numerous torture chambers for fellow human beings. These superstitions cause unimaginable damage not only in this life but also the next one. If you do not want to suffer without any purpose in hell, do not create one for yourself or your loved ones. This can be done by transforming your thoughts and mindsets.

Bon Voyage

Earth's heavens and hells would only have very recently become populated with plasma bodies associated or linked with homo sapiens -perhaps less than about two hundred thousand years ago. As Science advances, it will elucidate the general findings in the religious and metaphysical literature. Even precognition can be easily explained by the different temporal structures in succeeding universes. As discussed in chapter 11, the "present" moment of a higher energy universe covers both the past and future of a lower-energy universe. Hence, anyone operating from the physical-etheric body, which is linked to an ordinary matter body, would be both in the past and future of the ordinary matter reality. Plasma dynamics can easily explain telepathy since dark plasma bodies can communicate directly using dark electromagnetic waves, as discussed in chapter 10.

Columbus explored the Americas across the Atlantic Ocean less than a thousand years ago; we have plumbed the oceans and have sent missions to the Moon; and are preparing to go to Mars in the near future. With a greater understanding of dark matter and dark sectors, the stage is set for science to explore the vast dark plasmaspheres of Earth and our neighboring planets, which may host trillions upoun trillions of plasma and photonic-based lifeforms. Our familiar 3d world is only an infinitesimally small island in this unimaginably vast multi-dimensional multiverse. Certainly, our personal life journey within the local multiverse (i.e., samsara) could never have been imagined to be so challenging and at the same time so exciting. It is clear that your journey does not end with the death of your ordinary matter body and extends to many different planets in different universes at longer time-scales.

Bon Voyage!

Play the [Virtual] Game [of]
Existence to the End
Of the Beginning
John Lennon
Tomorrow Never Knows

Glossary

Aura: Radiation from plasma bodies.

Axions: Axions are light, low-mass, slow-moving hypothetical elementary dark matter particles that don't have a charge. They interact weakly with ordinary matter. In theory, axions would decay into a pair of ordinary light particles (photons) which could be detected.

Birkeland Currents: Electric currents that flow along magnetic field lines, which act like wires guiding the current in circuits. Acupuncture meridians and microcosmic orbits in Chinese acupuncture, nadis in Yogic literature, and channels in Tibetan Buddhism, consist of Birkeland currents, as well as currents within double-layers.

Chakras: Vortexes within plasma bodies, caused by magnetohydrodynamics. The term is primarily used in Hindu metaphysics. Other similar terms: Wheels (Tibetan Buddhism.)

Complex Plasma: Complex (or dusty) plasma contains charged dust particles mixed in with electrons and positive ions. The electrons are so tiny compared to the dust that they easily stick to the dust particles' surfaces. When this happens, the dust particles accumulate a large number of negative charges, increasing their electrostatic potential. This causes interesting nonlinear dynamics, which result in intricate processes and interesting features in complex plasma.

Dark Ionization: This is a process which occurs when there is a high density of dark matter particles interpenetrating a volume where there is a high density of ordinary matter particles. In this scenario, there is a higher probability that dark matter particles will collide with ordinary matter particles, sending ordinary electrons into higher energy states and subsequently emitting ordinary photons as the electrons return to their original states, releasing light and heat. Alternatively, if the collision force is greater, these electrons may become displaced from atoms. This would ionise the matter to produce ordinary plasma which would be visible temporarily, and also generate a current flow due to the free electrons displaced. (This term was minted by the author.)

Dark (or Invisible) Matter: This is invisible matter, which scientists believe make-up about 85 per cent of the matter of our universe. It interacts weakly with ordinary matter, which makes up the remaining 15 per cent, and consists of exotic particles, unless otherwise stated. (Exotic particles are particles that are not currently found in the standard model of particle physics.)

Dark Plasma: This is a plasma (see below) composed of self-interacting dark matter particles, which the author currently estimates to be about 15 per cent of all matter in the universe.

Dark Plasma Bodies: These are the biological bodies of life-forms which are composed of dark plasma. Other similar terms used in the metaphysical literature: Subtle, physical-etheric, astral, mental, spiritual, and causal bodies.

Dark Plasmaspheres/Plasmaverses: These are environments composed largely of dark plasma, including dark plasma planets and universes (or "plasmaverses.")

Dark Photon: In mainstream science, the dark photon is a hypothetical dark sector particle, proposed as a force carrier for dark electromagnetism. It can convert to ordinary photons, but unlike the latter, it has rest mass. In this book, however, the dark photon is hypothesized to represent a class of highly energetic dark matter particles which are massless in the vacuum and gain mass in dark plasma. The huge energy density of these dark photons warp spacetime, generating shadow gravity that shapes the distribution of matter in lower energy universes, including ours. It will allow for kinetic mixing, i.e., it will oscillate, converting to ordinary photons and back again.

Dark Photonics: This is the study and application of photonics (see below) in the dark sectors, universes, or branes.

Dark Sector: In particle physics, the dark sector (sometimes also known as the "hidden sector,") is a hypothetical collection of yet-unobserved quantum fields and their corresponding hypothetical particles and forces, in a dimensional structure that may be different from the ordinary matter universe. To this extent, they can be considered separate dark universes. The interactions between the hidden dark sector particles and standard model particles are weak, indirect, and typically mediated through gravity or other theorized exotic particles.

Eternal Inflation: Eternal inflation is a theoretical inflationary universe model, which is an extension of the big bang theory. The inflation could be eternal, leading to a multiverse in which the high energy vacuum condenses into lower energy bubble universes, whose properties, including fundamental constants and dimensional structure, differ from one bubble universe to another.

Fine Structure Constant: This is a dimensionless constant that measures the strength of the electromagnetic interaction between charged particles.

Imaginal Realm: This realm is in between dreams and reality, between mind and matter. The term was coined by Henry Corbin, a theologian and Sufi mystic.

Karma: The term "karma" is used in many different ways by various religions. Essentially, it refers to a principle or law of moral causation. In this book, it is simply taken to refer to the nature and composition of the relevant bodies of the conscious agent, as a consequence of past actions, mental and emotional states. The body, the temporary vehicle for non-local consciousness to observe through a local frame of reference, determines the future worldline of the conscious agent, including the corresponding state of consciousness and the universe it will inhabit. A high density and low-energy dark plasma body generates a worldline for the conscious agent that becomes increasingly entangled with matter and spacetime. A low density and high-energy dark plasma body generates a worldline that becomes increasingly disentangled from matter and spacetime.

Lambda (Greek Λ) - CDM (Cold Dark Matter) Model: This model expresses the big bang cosmological theory with measurable parameters in a universe containing three major components: first, a cosmological constant denoted by Lambda (Greek Λ) associated with dark energy; second, the postulated cold dark matter (abbreviated CDM); and third, ordinary matter.

Loop quantum gravity (LQG): This is a theory to merge quantum mechanics and general relativity, incorporating the standard model of particle physics. It is based directly on Einstein's geometric formulation rather than the treatment of gravity as a force. It postulates that the structure of space and time is composed of finite loops woven into an extremely fine fabric or network.

Plasma: This is the fourth state of matter. It can be generated when heat (or other energy) is added to atoms, causing the electrons to gain enough energy to break free from the atom. The soup of non-neutral atoms and freed electrons is said to be "ionized," and is a plasma. In other words, it is non-atomic matter.

Photonics: The study of light and other types of radiant energy, whose quantum unit is the photon, and related technologies. These include lasers, optical fibers, lenses, optical sensors, photonic propulsion, quantum computing, and data storage, among others. A subset of photonics is plasmonics, which takes advantage of the coupling of light to electric charges. The plasmon is a quantum unit of plasma.

Magnetohydrodynamics (or "MHD" for short): A study of the magnetic properties and behaviour of electrically conducting fluids. Examples of such magnetofluids include plasmas, liquid metals, salt water, and electrolytes.

Mirror or Shadow Matter: Mirror matter, also called shadow matter or Alice matter, is a hypothetical counterpart to ordinary matter. Mirror matter is self-interacting and is currently thought to use the weak force to interact with ordinary matter, other than gravity. They are synonymous with dark matter.

NDE: An NDE, which stands for "near-death-experience," is a distinct subjective experience that a minority of people report after a near-death episode, where a person is either clinically dead, near death, or in a situation where death is likely or expected.

Ordinary matter: Matter which is composed of standard particles (see below.)

Ovoid: Oval-shaped, three-dimensional object, with one end more sharply pointed than the other (like a chicken's egg.)

Scalar Field Dark Matter (SFDM): This is a classical, minimally coupled, scalar field postulated to account for the bulk of dark matter. A scalar filed contains only a magnitude (and no direction) at each point in space, for e.g., fields that depict temperature or pressure distribution.

Shell (3d) or Ring (2d): A spherical layer within a plasmasphere.

Shadow Gravity: This is gravity emanating from higher dimensional dark sectors, which interact with ordinary matter.

Standard Model of Particle Physics: The theory describing the electromagnetic, weak, and strong interactions in the universe as well as classifying all known elementary particles.

Standard Particles: As used in the terminology in this book, the particles that are included in the current Standard Model of Particle Physics and have been detected. Otherwise, they are identified as "exotic particles."

Super Particles: As used in the terminology in this book, the particles that are currently not included in the current Standard Model of Particle Physics and have not been detected. These include theorized "exotic particles" in science, and the "super-physical particles" discussed in metaphysical literature.

Superstring Theory: Superstring theory explains all the particles and fundamental forces of nature in one theory by modeling them as vibrations of tiny supersymmetric strings.

Supersymmetry: Supersymmetry theory predicts that for every force-carrying particle in the Standard model of particle physics, there is a matter particle and vice versa. None of these particles have been detected up to today. They are candidates for dark matter particles.

UFO/UAP: An unidentified flying object (UFO) or an unidentified aerial phenomena (UAP) is an object or light seen in the sky, the appearance and/ or flight dynamics of which do not suggest a logical, conventional flying object and which remains unidentified after close scrutiny of all available evidence by persons who have the technical expertise. Biological UFOs are believed to be unidentified flying life-forms.

References

Introduction

Stephen Hawking, *A Brief History of Time*, pp. 51, Bantam Books, 1995, 1988.

Part I: Heavens and Hells in History

Chapter 1: What's the Weather in Heaven Today?

Leadbeater Charles, W., *Inner Life*, Madras (India,) The Theosophical Publishing House, 1910-11.

Kaji, Hiralal. *The Great Mystery of Life Beyond Death*. New Age Books, 2003.

The Katha Upanishad (Sacred Wisdom Scriptures,) Ambikananda Saraswati (Translator,) Frances Lincoln Ltd, 2001. Historical Hindu Scriptures.

The Christian Bible, RSV.

Wright, Edward, J. *The Early History of Heaven*, Oxford University Press, New York, 2000.

The Holy Quran.

Chapter 2: Where the Hell Am I?

Fernandes, Joaquim and D'Armada, Fina. *Heavenly Lights: The Apparitions at Fatima and the UFO Phenomenon*, Eccenova Publications, forthcoming title in 2005.

Bart D. Ehrman, *Heaven and Hell: A History of the Afterlife*, Simon & Schuster, 2020.

The Christian Bible, RSV.

Chapter 3: Quick Tour of the Earth

National Geographic, *Earth Structure*, National Geographic Website.

NASA, *Earth's Atmospheric Layers*, 2013, 2017. NASA Science, *Inside the Moon*, 2022. NASA Website.

Part II: Dark Matter

Chapter 4: Can Someone Turn-On the Lights?

Siegfried, Tom, *Strange Matters*, Berkeley Publishing Group (a division of Penguin Group,) 2002.

Rubin, Vera C. 1983. *Dark Matter in Spiral Galaxies*. Scientific American 248, no. 6 (June 1983,) 96–109.

Robert Minchin, Jonathan Davies, Michael Disney, Peter Boyce, Diego Garcia, Christine Jordan, Virginia Kilborn, et al. 2005. *A Dark Hydrogen Cloud in the Virgo Cluster*. ArXiv: astro-ph/0502312, 2005.

Peat, F. David, *Superstring and the Search for the Theory of Everything*, 1988
Einstein, A., and W. De Sitter. 1932. *On the Relation between the Expansion and the Mean Density of the Universe*. Proceedings of the National Academy of Sciences, Number 3, March 15, 1932.
Morris, Richard, *The Edges of Science*, 1990.
Goldsmith, Donald, *The Runaway Universe*, 2000.
Blavatsky, H. P., *The Secret Doctrine*, Madras (India,) The Theosophical Publishing House, 1905.
Leadbeater, Charles W., *The Chakras*, Madras (India,) The Theosophical Publishing House, 1927.
Brennan, Barbara A., *Hands of Light*, Bantam Books, 1987.
Pearsall, Paul, *The Heart Code*, HarperCollins, 1998.
Mckee, Maggie, *Dark Matter Clouds May Float Through Earth*, New Scientist.com news service, 26 January 2005.
Dark-Matter Highway may be Streaming Through the Earth, Press release by Rensselaer Polytechnic Institute, March 24, 2004.

Chapter 5: What Exactly is Dark Matter?

Kane, Gordon. 2000. *Supersymmetry*. Basic Books, New York. 2000.
Peat, F. David, *Superstring and the Search for the Theory of Everything*, 1988.
Siegfried, Tom, *Strange Matters*, Berkeley Publishing Group (a division of Penguin Group,) 2002.
Spergel, David N., and Paul J. Steinhardt. 2000. "Observational Evidence for Self-Interacting Cold Dark Matter." Phys.Rev.Lett.84:3760-3763, 2000. arXiv: astro-ph/9909386, 2000.
Gribbin, John, *In Search of SUSY*, 1998.
Taimni, I. K., *Science and Occultism*, Madras (India,) The Theosophical Publishing House, 1974.
Leadbeater, Charles, W., *Inner Life*, Madras (India,) The Theosophical Publishing House, 1910-11.
Leadbeater, Charles, W., *Textbook of Theosophy*, Madras (India,) The Theosophical Publishing House, 1914.
Leadbeater, Charles W., *Some Glimpses of Occultism*, Madras (India,) The Theosophical Publishing House, 1913.
Besant, Annie and Leadbeater, Charles W., *Occult Chemistry*, Madras (India,) The Theosophical Publishing House, 1919.
Besant, Annie, *Man and His Bodies*, Madras (India,) The Theosophical Publishing House, 1952. First published in 1896.

Chapter 6: Shadow and Mirror Universes

Morris, Richard, *The Edges of Science*, 1990.
Gribbin, John, *In Search of SUSY*, 1998.
Peat, F. David, *Superstring and the Search for the Theory of Everything*, 1988
Silagadze, Z. K., *Mirror objects in the Solar System?* 2001.
Foot, Robert, *Shadowlands*, 2002.
Hitchcock, J., *The Web of the Universe*, 1991.
Besant, Annie, *Death and After*, Madras (India,) The Theosophical Publishing House, 1893.
Powell, Arthur, E., *The Astral Body*, Madras (India,) The Theosophical Publishing House, 1927.
Monroe, Robert, A., *Journeys Out of the Body*, 1972.

Tiller, William, A., *Science and Human Transformation*, Pavior Publishing, California, 1997.
Hutchison, Michael, *Megabrain*, Ballantine Books, New York, 1986, 1991.
Becker, O. Robert and Selden Gary, *The Body Electric*, William Morrow and Company, Inc., New York, 1985.
Brennan, Barbara, A., *Hands of Light*, Bantam Books, 1987.

Part III: Plasma

Chapter 7: Our Plasma Universe

Klaus Dolag, Matthias Bartelmann and Harald Lesch, *SPH simulations of magnetic fields in galaxy clusters, astronomy & astrophysics*, 1999.
Peratt, Anthony, L. Los Alamos National Laboratory. *The Evidence for Electrical Currents in Cosmic Plasma*. No date given.
Corradi, R. L. M., Sanchez-Blazquez, P., Mellema G., Gianmanco, C., Schwarz, H. E., *Rings in the Haloes of Planetary Nebulae*. December 15, 2005.

Chapter 8: Dark Plasmaspheres

Arabadjis, J. S., Bautz, M. W., Mass Profiles of Galaxy Cluster Cores, 16 February 2005.
Alfred, Jay. 2005. *Our Invisible Bodies*. Trafford Publishing, Bloomington, Indiana, USA. 2006.
Leadbeater, Charles, W., *Inner Life*, Madras (India,) The Theosophical Publishing House, 1910-11.
Leadbeater, Charles, W., The Chakras, Madras (India,) The Theosophical Publishing House, 1927.
Phil Schewe, Ben Stein, *AIP Bulletin of Physics News*, Number 705 #1, October 20, 2004.
Corradi, R. L. M., Sanchez-Blazquez, P., Mellema G., Gianmanco, C., Schwarz, H. E., *Rings in the Haloes of Planetary Nebulae*, December 15, 2005.
H. Thomas, G. E. Morfill, and V. Demmel, J. Goreet, B. Feuerbacher and D. Mohlmann. 1994. *Plasma Crystal: Coulomb Crystallization in a Dusty Plasma*, Physical Review Letters, Vol. 73, No.5, 1994.
Whitton, Joel and Fisher, Joe, *Life between Life*, Doubleday, 1986.
George Amarandei, Cezar Gaman, Dan G. Dimitriu, Codrina Ionita, Mircea Sanduloviciu, et al. *Similarities between the generation and dynamics of concentric and non-concentric multiple double layers*. 2004. hal-00001963.
Peratt, Anthony, L. Los Alamos National Laboratory. *The Evidence for Electrical Currents in Cosmic Plasma*. No date given.
Yogananda, Paramahansa, *Autobiography of a Yogi*. Self-realization Fellowship, Los Angeles, 1946.
Saying attributed to Jesus of Nazareth, John 14:2, The New Testament, *The Christian Bible, RSV*.
Leadbeater, Charles W., *Astral Plane*, Madras (India,) The Theosophical Publishing House. 1910.
Moody, Raymond, A. *Life after Life*. Harper Collins, San Francisco, 2001. First published in 1975.
Cruz, Joan Carroll. *Mysteries, Marvels and Miracles in the Lives of the Saints*. Tan Books and Publishers, 1997.
John G Cramer. *Falling Through the Pelucidar*. Web article.
Siegfried, Tom, *Strange Matters*. Berkeley Publishing Group (a division of Penguin Group, 2002.

Chapter 9: Dark Plasma Bodies *Biology and Appearance*

Cohen, David, *Plasma Blobs Hint at New Life*, New Scientist, September 2003.

Alfred, Jay. 2005. *Our Invisible Bodies.* Trafford Publishing, Bloomington, Indiana, USA. 2006.

Leadbeater, Charles, W., *Inner Life*, Madras (India,) The Theosophical Publishing House, 1910-11.

Leadbeater, Charles, W., *Textbook of Theosophy*, Madras (India,) The Theosophical Publishing House, 1914.

Greyson, Bruce. *NDE research*: University of Virginia, Division of Perceptual Studies.

NDERF (Near-Death Experience Research Foundation), www.nderf.org – various studies.

Chapter 10: Dark Plasma Bodies *Communications and Transport*

Alfred, Jay. 2005. *Our Invisible Bodies.* Trafford Publishing, Bloomington, Indiana, USA. 2006.

Leadbeater, Charles, W., *Inner Life*, Madras (India,) The Theosophical Publishing House, 1910-11.

Peratt, Anthony, L. Los Alamos National Laboratory. *The Evidence for Electrical Currents in Cosmic Plasma.* No date found.

Part IV: The Local Multiverse

Chapter 11: M-Theory and Eternal Inflation

Randall, Lisa, *Warped Passages*, HarperCollins Publishers, New York, 2005.

Andrei, Linde. 2005. *Particle Physics and Inflationary Cosmology.* Department of Physics, Stanford University. arXiv:hep-ph/0503203, 2005.

Vilenkin, Alexander. 2006. *Many Worlds in One - The Search for Other Universes.* Hill and Wang, New York. 2006.

Dark Matter" forms Ghost Universe That Mirrors Our Own, New Theory Shows. University of California release, 10 November 2003.

Yogananda, Paramahansa, *Autobiography of a Yogi.* Self-realization Fellowship, Los Angeles, 1946.

Leadbeater, Charles, W., *Inner Life*, Madras (India,) The Theosophical Publishing House, 1910-11.

Alfred, Jay. 2005. *Our Invisible Bodies.* Trafford Publishing, Bloomington, Indiana, USA. 2006.

Chapter 12: Bodies and their Universes

Judith Hopper and Dick Teresi. *The Three-Pound Universe.* MacMillan, 1986.

Applied Supersymmetry and Quantum Gravity II, Proceedings of the Workshop at Imperial College, London 5-10 July 1996, World Scientific Publishing, 1997.

Vilenkin, Alexander. 2006. *Many Worlds in One - The Search for Other Universes.* Hill and Wang, New York. 2006.

Randall, Lisa, *Warped Passages*, HarperCollins Publishers, New York, 2005.

Ofer Aharony, Steven S. Gubser, Juan Maldacena, Hirosi Ooguri, and Yaron Oz, *Large N Field Theories, String Theory and Gravity*, 1999.

Beckenstein, Jacob, D., *Information in the Holographic Universe.* Scientific American feature article, August 2003.

Duff, Michael J., *Supermembranes: An Introduction, Gauge Theories, Applied Supersymmetry and Quantum Gravity II*, Proceedings of the Workshop at Imperial College, London 5 — 10 July 1996, World Scientific Publishing, 1997.

Alfred, Jay. 2005. *Brains and Realities*. Trafford Publishing, Bloomington, Indiana, USA. 2006.
Alfred, Jay. 2005. *Our Invisible Bodies*. Trafford Publishing, Bloomington, Indiana, USA. 2006.
Besant, Annie, *Death and After*, Madras (India,) The Theosophical Publishing House, 1893.
Leadbeater, Charles, W., *Inner Life*, Madras (India,) The Theosophical Publishing House, 1910-11.
Leadbeater, Charles, W., *The Hidden Side of Things*, Madras (India,) The Theosophical Publishing House, 1913.

Part V: Dark Earths

Chapter 13: Dark Earths

Alfred, Jay. 2005. *Our Invisible Bodies*. Trafford Publishing, Bloomington, Indiana, USA. 2006.
Leadbeater, Charles, W., *Textbook of Theosophy*, Madras (India,) The Theosophical Publishing House, 1914.
Leadbeater, Charles, W., *Inner Life*, Madras (India,) The Theosophical Publishing House, 1910-11.
Allan Barton, *States of Matter-States of Mind*, IOP Publishing Ltd., 1997.

Chapter 14: Classical Hells

Jung, Carl, *Memories, Dreams, Reflections: An Autobiography*. Pantheon Books, 1957.
Bart D. Ehrman, *Heaven and Hell: A History of the Afterlife*, Simon & Schuster, 2020.
Alfred, Jay. 2005. *Our Invisible Bodies*. Trafford Publishing, Bloomington, Indiana, USA. 2006.
Dalai Lama, Bstan- "dzin-rgya-mtsho. *Sleeping, Dreaming and Dying*. Francisco J. Varela (editor,) Wisdom Publications, 1999.
Yogananda, Paramahansa, *Autobiography of a Yogi*. Self-realization Fellowship, Los Angeles, 1946.
Huxley, Aldous. *The Doors of Perception and Heaven and Hell*, 1954, 1956, Harper & Brothers (US); Chatto & Windus (UK).
Nielsen, T. A. (1993). Changes in the kinesthetic content of dreams following somatosensory stimulation of leg muscles during REM sleep. Dreaming, 3(2), 99-113.
Swedenborg, Emanuel, *Heaven and Hell*, first published in the 1700s.
Leadbeater, Charles, W., *Inner Life*, Madras (India,) The Theosophical Publishing House, 1910-11.
Ring, Kenneth, *The Omega Project: Near- death experiences, UFO encounters and the Mind at Large*, New York, William Morrow, 1993.
Greyson, Bruce. *NDE research*: University of Virginia, Division of Perceptual Studies.
NDERF (Near-Death Experience Research Foundation), www.nderf.org – various studies.

Chapter 15: Classical Heavens

Alfred, Jay. 2005. *Our Invisible Bodies*. Trafford Publishing, Bloomington, Indiana, USA. 2006.
Dalai Lama, Bstan- "dzin-rgya-mtsho. *Sleeping, Dreaming and Dying*. Francisco J. Varela (editor,) Wisdom Publications, 1999.
Swedenborg, Emanuel, *Heaven and Hell*, first published in the 1700s.
Ring, Kenneth, *The Omega Project: Near- death experiences, UFO encounters and the Mind at Large*, New York, William Morrow, 1993.
Bart D. Ehrman, *Heaven and Hell: A History of the Afterlife*, Simon & Schuster, 2020.

Chapter 16: Semi-Classical Heavens

Alfred, Jay. 2005. *Our Invisible Bodies.* Trafford Publishing, Bloomington, Indiana, USA. 2006.
Martin Gardner, *The Ambidextrous Universe.* Basic Books, 1964.
M. A. Prakapenia and G. V. Vereshchagin. *Bose-Einstein Condensation in Relativistic Plasma*, 30 January 2020, EPLA, 2020.
Zijun, Yan, *General Thermal Wavelength, and its Applications.* Eur. J. Phys. 21 (2000) pp. 625-631.
P. Michel. *Plasma Photonics: Manipulating Light Using Plasmas* (LDRD full report.) 2021.
Monroe, Robert A., *Ultimate Journey.* Doubleday, New York, 1994.
Swedenborg, Emanuel, *Heaven and Hell,* first published in the 1700s. Krieger, Kim, Focus Physical Review, 2003.
Swedenborg, Emanuel, *The Spiritual diary: records and Notes Made by Emanuel Swedenborg Between 1746 and 1765 from His experiences in the Spiritual World*: Sections 1539-3240 v.2, Stephen McNeilly (editor,) G. Bush (Translator,) J.H. Smithson (Translator,) The Swedenborg Society, 2002.
Larsen, Stephen Manuel, Emanuel Swedenborg: *The Universal Human and Soul-Body Interaction*, The Classics of Western Spirituality, New York, Paulist Press, 1984.
Yogananda, Paramahansa, *Autobiography of a Yogi.* Self-realization Fellowship, Los Angeles, 1946.
Leadbeater, Charles, W., *Inner Life*, Madras (India,) The Theosophical Publishing House, 1910-11.
Powell, Arthur, E., *The Mental Body.* Madras (India,) The Theosophical Publishing House, 1927.

Chapter 17: Quantum Realities

Alfred, Jay. 2005. *Our Invisible Bodies.* Trafford Publishing, Bloomington, Indiana, USA. 2006.
Krippner, S., and Ruhin, D., The Energies of Consciousness, New York, Gordon and Breach, 1975.
Yogananda, Paramahansa, *Autobiography of a Yogi*, Self-realization Fellowship, Los Angeles, 1946.
Buddhist Scriptures, Pali Canon, D.N. (Digha Nikaya) 1.213, Udana 8.1 (Nibbana Sutta), M.L.S. (Middle Length Sayings, Pali Text Society Translation Series) 3.93-4.
Bohm, David and Hiley, Basil, *The Undivided Universe.* Routledge, 1995.
McTaggart, Lynne, *The Field*, HarperCollins, 2003.

Part VI: Inter-Spheric Interactions

Chapter 18: Dark Plasma UFOs

Alfred, Jay. 2005. *Our Invisible Bodies.* Trafford Publishing, Bloomington, Indiana, USA. 2006.
Taimni, I. K., *Science and Occultism*, Madras (India,) The Theosophical Publishing House, 1974.
Ring, Kenneth, *Toward an Imaginal Interpretation of "UFO Abductions.* MUFON Journal number 253, pp. 3-9, May 1989.
Ring, Kenneth, *The Omega Project: Near- death experiences, UFO encounters and the Mind at Large*, New York, William Morrow, 1993.
Ring, Kenneth, *Near-death and UFO encounters as Shamanic Initiations: Some Conceptions and Evolutionary Implications*, Vol. 11, No. 3, Winter 1989, Web article.
Jacques Vallee. Dimensions: *A Casebook of Alien Contact.* Ballantine Books, 1989.

Chapter 19: Plasma Deities and Angels

Alfred, Jay. 2005. *Our Invisible Bodies.* Trafford Publishing, Bloomington, Indiana, USA. 2006.

Talbot, Michael, *The Holographic Universe,* HarperCollins, Great Britain, 1991.

Fernandes, Joaquim and D'Armada, Fina. *Heavenly Lights: The Apparitions at Fatima and the UFO Phenomenon,* Eccenova Publications, forthcoming title in 2005.

Jacques Vallee. Dimensions: *A Casebook of Alien Contact.* Ballantine Books, 1989.

William, R., Corlis, *Handbook of Unusual Natural Phenomena,* Gramercy Books, New York, 1977, 1983.

Yogananda, Paramahansa, *Autobiography of a Yogi,* Self-realization Fellowship, Los Angeles, 1946.

Chapter 20: Dark Plasma Ghosts

Alfred, Jay. 2005. *Our Invisible Bodies.* Trafford Publishing, Bloomington, Indiana, USA. 2006.

Leadbeater, Charles W., *Invisible Helpers,* Madras (India,) The Theosophical Publishing House, 1911.

Leadbeater, Charles, W., *Inner Life,* Madras (India,) The Theosophical Publishing House, 1910-11.

Cowen, Ron. *A Cosmic Crisis? Dark Doings in the Universe,* Science News Online, Oct 2001.

McManus, Craig, *Everything You Ever Wanted to Know about Ghosts, (But Were Afraid to Ask.)* Web article.

Oester, Dave, *The Thermal Scanner,* Web article.

H. Evans, *Seeing Ghosts - Experiences of the Paranormal,* John Murray, London, 2002.

Part VII: Moving Out

Chapter 21: Beyond Earth

Alfred, Jay. 2005. *Our Invisible Bodies.* Trafford Publishing, Bloomington, Indiana, USA. 2006.

Klypin et al. 1999. *Where Are the Missing Galactic Satellites?* The Astrophysical Journal 522 (1999): 82-92.

Foot, Robert, Silagadze, Z. K., *Do mirror planets exist in our solar system?* arXiv: astro-ph/0104251v1 15 Apr 2001.

Pearson, E., Norman. *Space, Time and Self,* The Theosophical Publishing House, first published in 1957.

Leadbeater, Charles, W., *Inner Life,* Madras (India,) The Theosophical Publishing House, 1910-11.

Yogananda, Paramahansa, *Autobiography of a Yogi,* Self-realization Fellowship, Los Angeles, 1946.

Printed in the United States
by Baker & Taylor Publisher Services